Sustainability Sciences in Asia and Africa

# Sustainable Agriculture and Food Security

**Series Editor**

Rajeev K. Varshney, Centre for Crop and Food Innovation, Food Futures Institute
Murdoch University
Perth, WA, Australia

Sustainability as a scientific endeavor requires broad understanding of the interconnections in our global environment. It is therefore positioned at the crossroads of research in natural sciences, technology and human sciences. This book series supports the global efforts by providing timely coverage of the progress, opportunities, and challenges towards achieving sustainable development goals (SDGs) in Asia and Africa. It fosters the research in transdisciplinary academic fields spanning across agriculture and global food security, health and diseases, access to clean water and affordable energy, climate resilience and effects of global climate change on biodiversity, water and land. This series is planned to narrate the stories of scientific development and progression towards SDGs from the regions of Asia and Africa. The overall aim of the series is to provide a comprehensive publication platform and act as a knowledge engine in the growth of sustainability sciences with a special focus on developing nations. The series constitute of four sub-series each focusing on a specific theme closely linked to a set of SDGs. Each sub-series is leaders and experts in agricultural sciences, environmental sciences, biomedicine and medical sciences respectively. Books in this series are useful for environmentalists, biotechnologists, plant scientists, medical practitioners, conservation biologists, policy makers, NGOs and scientific bodies working in the areas of sustainability sciences in the developing regions.

This book series support the global efforts towards sustainability by providing timely coverage of the progress, opportunities, and challenges of sustainable food production and consumption in Asia and Africa. The series narrates the success stories and research endeavors from the regions of Africa and Asia on issues relating to SDG 2: Zero hunger. It fosters the research in transdisciplinary academic fields spanning across sustainable agriculture systems and practices, post- harvest and food supply chains. It will also focus on breeding programs for resilient crops, efficiency in crop cycle, various factors of food security, as well as improving nutrition and curbing hunger and malnutrition. The focus of the series is to provide a comprehensive publication platform and act as a knowledge engine in the growth of sustainability sciences with a special focus on developing nations. The series will publish mainly edited volumes but some authored volumes. These volumes will have chapters from eminent personalities in their area of research from different parts of the world.

Ashok K. Patra
Editor

# Nanofertilizers in Agriculture

Springer

*Editor*
Ashok K. Patra
ICAR - Indian Institute of Soil Science
Bhopal, Madhya Pradesh, India

ISSN 2730-6771 ISSN 2730-678X (electronic)
Sustainability Sciences in Asia and Africa
ISSN 2730-6798 ISSN 2730-6801 (electronic)
Sustainable Agriculture and Food Security
ISBN 978-981-97-8111-9 ISBN 978-981-97-8112-6 (eBook)
https://doi.org/10.1007/978-981-97-8112-6



This Springer imprint is published by the registered company Springer Nature Singapore Pte Ltd.
The registered company address is: 152 Beach Road, #21-01/04 Gateway East, Singapore 189721, Singapore

# Foreword

The genesis of nanofertilizers stems from the intersection of nanotechnology and agriculture, where scientists recognized the immense potential of nanomaterials to revolutionize conventional fertilizer systems. Through meticulous research and innovation, nanofertilizers have emerged as efficient carriers of nutrients, capable of enhancing nutrient uptake, minimizing losses, and improving crop yields. Their nano-sized particles facilitate targeted delivery, ensuring nutrients reach plants precisely when and where they are needed, thereby mitigating environmental impacts and optimizing resource utilization.

Recent developments in nanofertilizers have showcased remarkable strides, ranging from tailored nutrient release mechanisms to multifunctional formulations that address specific crop requirements and soil conditions. Nanoencapsulation techniques, functionalized nanomaterials, and nanosensors have expanded the scope of nanofertilizers, promising customized solutions for diverse agricultural contexts.

However, amidst the optimism surrounding nanofertilizers, significant constraints and challenges persist. Concerns regarding nanoparticle toxicity, environmental impacts, regulatory frameworks, and scalability pose formidable hurdles to widespread adoption. Addressing these challenges demands interdisciplinary collaboration, rigorous risk assessments, and transparent communication to ensure the safe and responsible integration of nanofertilizers into agricultural systems.

Looking ahead, the research efforts must prioritize understanding the fate and behavior of nanoparticles in soil-plant systems, optimizing formulations for different crops and soil types, and assessing long-term environmental implications. Furthermore, concerted efforts are needed to foster stakeholder engagement, facilitate knowledge exchange, and promote evidence-based policymaking to navigate the complex landscape of nanotechnology in agriculture.

The relevance of "Nanofertilizers—An Emerging Technology for Sustainable Agriculture" extends far beyond scientific circles, resonating with policymakers, industry stakeholders, farmers, and environmental advocates alike. By offering a comprehensive exploration of the promise, challenges, and implications of nanofertilizers, this book serves as a beacon of insight and guidance for navigating the transformative potential of nanotechnology in agriculture, ultimately contributing to the pursuit of sustainable and resilient food systems on a global scale.

I congratulate and extend my appreciation to Dr. Ashok K. Patra for his efforts for editing and bringing out this book.

Secretary, DARE & DG, ICAR Himanshu Pathak
Krishi Bhawan, New Delhi, India

सत्यमेव जयते

डॉ. हिमांशु पाठक
**DR. HIMANSHU PATHAK**
सचिव (डेयर) एवं महानिदेशक (आईसीएआर)
Secretary (DARE) &
Director General (ICAR)

भारत सरकार
कृषि अनुसंधान और शिक्षा विभाग एवं
भारतीय कृषि अनुसंधान परिषद
कृषि एवं किसान कल्याण मंत्रालय, कृषि भवन, नई दिल्ली–110 001

**GOVERNMENT OF INDIA**
**DEPARTMENT OF AGRICULTURAL RESEARCH AND EDUCATION (DARE)**
**AND**
**INDIAN COUNCIL OF AGRICULTURAL RESEARCH (ICAR)**
**MINISTRY OF AGRICULTURE AND FARMERS WELFARE**
**Krishi Bhavan, New Delhi 110 001**
Tel: 23382629 / 23386711 Fax: 91-11-23384773
E-mail: dg.icar@nic.in

# FOREWORD

The transformative potential of nanofertilizers in agriculture offers innovative approach to revolutionize the conventional fertilizer practices in agriculture. Harnessing the unique properties of nanoscale materials, nanofertilizers provide advanced nutrient delivery systems that enhance nutrient use efficiency, reduce losses, and contribute to sustainable crop productivity. By ensuring precise nutrient delivery at the right time and location, they optimize resource use while significantly mitigating environmental impacts.

The innovative strategies such as nanoencapsulation and functionalized nanoparticles are indeed promising for optimizing nutrient delivery and reducing environmental impacts. Moreover, the nanosensors are revolutionizing nutrient monitoring, paving the way for real-time precision agriculture. These innovations demonstrate the adaptability of nanofertilizers across diverse agro ecological contexts. However, the road to integrating nanofertilizers into mainstream agriculture is fraught with challenges. Concerns over nanoparticle toxicity, environmental risks, regulatory compliance, and the scalability of production remain critical barriers. Addressing these challenges necessitates collaborative efforts among scientists, policymakers, industry, and other stakeholders. Rigorous research, risk assessments, and transparent, evidence-based policies will be essential to ensure the safe and sustainable application of this technology. Moving forward, research priorities must include understanding the complex interactions of nanoparticles in soil-plant systems, optimizing formulations for diverse agronomic conditions, and assessing long-term environmental impacts. Equally vital is fostering dialogue and cooperation among stakeholders to bridge knowledge gaps, promote public trust, and drive informed policy-making.

*The book 'Nanofertilizers in Agriculture'* is a comprehensive work that underscores the importance in guiding future advancements in the field. This comprehensive work is set to influence research and decision-making for sustainable agriculture, fostering innovation that will benefit both food security and environmental sustainability. I appreciate the contribution of Dr. Ashok K. Patra in bringing out this publication, which will shape the future research and decision-making for sustainable agricultural practices.

**(Himanshu Pathak)**

**Dated the 28th November, 2024**
**New Delhi**

भारतीय कृषि अनुसंधान परिषद
कक्ष क्र. 101, कृषि अनुसंधान भवन-II, नई दिल्ली-110 012, भारत
**INDIAN COUNCIL OF AGRICULTURAL RESEARCH**
Room No. 101, Krishi Anusandhan Bhavan-II, Pusa, New Delhi-110012, India

**डॉ. सुरेश कुमार चौधरी**
उप महानिदेशक (प्राकृतिक संसाधन प्रबंधन)
**Dr. Suresh Kumar Chaudhari**
Deputy Director General (Natural Resources Management)

**26.11.2024**

# Foreword

Agriculture today faces the dual challenge of meeting growing food demands while preserving environmental sustainability. With rising populations and little scope for expansion of cultivable land coupled with impact of climate change, innovative approaches are essential to ensure food security and ecological balance. Nanofertilizers, as a transformative application of nanotechnology, offer promising solutions by enhancing nutrient use efficiency and addressing critical issues such as soil health, environmental degradation and mitigation of climate change.

This book, *Nanofertilizers in Agriculture*, edited by Dr. Ashok K. Patra, is a timely and comprehensive contribution to this evolving field. It provides a nuanced understanding of nanofertilizers, covering their development, mechanisms in soil and plants, and their impacts on agricultural and environmental systems. The insights presented here make it an indispensable resource for researchers, policymakers, and practitioners striving for sustainable agricultural advancements.

Nanofertilizers hold particular significance for Indian agriculture. As the country works to boost productivity while safeguarding natural resources, this technology offers innovative pathways to improve nutrient delivery systems, optimize resource use, and reduce environmental footprints. By examining applications in India and globally, this book highlights the potential of nanofertilizers to drive sustainable agricultural practices worldwide.

I commend Dr. Ashok K. Patra and the contributors for compiling this critical work. The knowledge and perspectives shared in this book will undoubtedly encourage further research, foster meaningful collaborations, and guide the development of resilient and sustainable agricultural production systems not only in the country but also in other parts of the world.

**(S.K. Chaudhari)**

Ph.: +91-11-25848364 Fax: +91-11-25848366 Email: ddg.nrm@icar.gov.in Website: www.icar.org.in

# Series Preface

The global food production systems require expanding their capacity to produce almost twice the current levels to safeguard the food security of the burgeoning population worldwide. More than 800 million people suffering from undernourishment worldwide pose a great risk to the attainment of Sustainable Development Goal (SDG) 2 of the UN, which targets "End hunger, achieve food security and improved nutrition and promote sustainable agriculture" within the next 7 years. The challenge is further exacerbated by the rising weather extremities, unpredictability in rainfall patterns, and pest–pathogen dynamics associated with global climate change, which have a profound negative impact on agricultural productivity and farm incomes worldwide. Also, the future targets of food production should be secured in resource-constrained agricultural settings and with the least environmental footprint, thus calling for sustainable innovations in agri-farming systems and enhanced participation of women in agriculture. The challenge to reduce hunger is alarming in developing nations, particularly Asia and Africa, which house the largest proportion of people suffering from malnutrition and other nutrition-related issues. Furthermore, the agri-food systems in Asia and Africa are severely constrained by the subsistence nature of farming, declining land and other agricultural resources, increasing environmental pollution, soil and biodiversity degradation, and climate change. Therefore, this book series, "Sustainable Agriculture and Food Security", has been planned to support the global efforts towards sustainability by providing timely coverage of the progress, opportunities, and challenges of sustainable food production and consumption in Asia and Africa. The series narrates the success stories and research endeavours from the regions of Africa and Asia on issues relating to SDG 2: Zero Hunger. It fosters research in transdisciplinary academic fields spanning sustainable agriculture systems and practices, post-harvest, and food supply chains. The focus of the series is to provide a comprehensive publication platform and act as a knowledge engine in the growth of sustainability sciences with a special focus on developing nations.

By 2050, the global population is projected to reach 9.7 billion people. Achieving food security for a growing population presents a challenge which is further exacerbated by climate change. Estimates suggest that global food production must increase by 70% by 2050 to meet the growing demand. To achieve higher yields, farmers often overuse chemical fertilizers, leading to soil, environmental, and water pollution. Additionally, the absorption efficiency of chemical fertilizers remains low, for example, 20–70% of applied nitrogen is lost to the environment, necessitating constant application and increasing production costs. To address these issues, innovative agricultural approaches are needed to ensure global food security while protecting ecosystems, biodiversity, and the climate. One such innovation is the use of "nanofertilizers", which employ nanotechnology to improve nutrient delivery efficiency, therefore reducing the amount of fertilizer required and minimizing environmental harm.

Nanofertilizers fall into three categories: (i) nanomaterials made of macronutrients, (ii) nanomaterials made of micronutrients, and (iii) nanomaterials serving as carriers for macronutrients. Due to their small particle size, nanofertilizers enhance plant absorption and retention capacity compared to chemical fertilizers. They can be tailored to specific crops and environmental conditions, ensuring nutrients are available when plants need them most. Unlike conventional fertilizers, which often result in nutrient waste and environmental harm, nanofertilizers release nutrients slowly and gradually into the soil. This reduces the chemical load on the soil, improves nutrient use efficiency, reduces adverse environmental effects, and decreases the frequency of application. In countries like India, nanofertilizers have already been adopted, and products such as nano-nitrogen, nano-zinc, and nano-copper are undergoing initial testing. Preliminary results have been promising, with maize, wheat, pearl millet, and mustard exhibiting significantly higher grain yields compared to conventional fertilizer controls. This suggests that nanofertilizers can reduce the need for nitrogen fertilizers while maintaining yield levels. Besides benefiting plant health, nanofertilizers also improve soil health by lowering chemical loads, increasing enzyme activity, improving microbial activity, and enhancing nutrient mobilization within the rhizosphere. However, there are risks associated with using nanofertilizers. Some studies indicate that nanofertilizers can negatively affect some nitrogen-fixing fungi, detritivores, and bacteria, reducing their growth. Furthermore, excessive quantities of nanofertilizers can harm crop, soil, and human health, necessitating careful assessment and regulatory frameworks before widespread commercial adoption.

In view of this, the present book, *Nanofertilizers in Agriculture*, edited by Ashok K. Patra, provides a platform to emphasize the potential of nanofertilizers in enhancing sustainability in agriculture and presents practical aspects of nanofertilizer adoption, including regulatory policies and cost-effectiveness. This book presents seven chapters written by experts and practitioners, each containing up-to-date and important insights into nanofertilizer technology. I would like to congratulate the editor, Ashok K. Patra, and all the authors for their valuable contributions. I am sure this book is a great resource for students, researchers, farmers, educators and extension agents, policymakers, and other stakeholders engaged in transforming agriculture around the world.

I wish to extend my sincere thanks and gratitude to the Springer staff, particularly Sanchi Bhimrajka, Associate Editor (Books, Medicine and Life Sciences), and

Naren Aggarwal, Editorial Director (Books, Asia Medicine and Life Sciences) for their constant support for the accomplishment of this compendium. The cooperation received from my supervisors and research leaders, including Prof. Andrew Deeks (Vice Chancellor), Prof. Peter Eastwood (Deputy Vice Chancellor—Research), and Prof. Peter Davies (Pro Vice Chancellor), and my laboratory colleagues, Rhys Copeland, Anu Chitikineni, and Vanika Garg from Murdoch University (Australia), is also gratefully acknowledged. I would like to thank my family members, Monika Varshney, Prakhar Varshney, and Preksha Varshney, for their love and support in discharging my duties as Series Editor.

Perth, WA, Australia Rajeev K. Varshney

# Preface

Among recent breakthroughs in agricultural sciences, nanofertilizers have emerged as one of the most promising technologies with the potential to revolutionize nutrient management and promote sustainable crop production. This book, *Nanofertilizers in Agriculture*, explores the potential, challenges, and future directions of this rapidly evolving field.

Nanofertilizers represent a paradigm shift in soil fertility management and plant nutrition. By leveraging nanotechnology, these advanced fertilizers enable precise nutrient delivery, significantly enhancing efficiency while reducing environmental losses. The unique physicochemical properties of nanoparticles—such as high surface area, controlled release mechanisms, and multifunctionality—facilitate targeted nutrient supply, optimizing plant uptake while mitigating adverse environmental impacts, including leaching and volatilization.

The development of nanofertilizers has been propelled by the need to address critical challenges in modern agriculture, including declining soil health, nutrient imbalances, and increasing uncertainties due to climate change. This book provides a comprehensive analysis of the mechanisms underlying nanofertilizer function, their applications, and their interactions with the environment. It also highlights innovative advancements in their formulation, such as the use of engineered nanomaterials and encapsulation techniques, which enhance their efficacy and adaptability across diverse agro-ecosystems.

While the potential benefits of nanofertilizers are immense, their widespread adoption is not without challenges. Issues related to nanoparticle safety, environmental persistence, scalability, and regulatory frameworks require careful evaluation. This book addresses these concerns, offering a balanced perspective on the scientific, ethical, and policy-related considerations essential for the safe and sustainable deployment of nanofertilizers in agriculture.

Bridging cutting-edge research with practical applications, this book brings together contributions from leading experts in the field. It caters to a diverse audience, including scientists, policymakers, industry professionals, and farmers, providing insights that are both scientifically rigorous and practically relevant. By presenting a holistic view of the opportunities and challenges associated with nanofertilizers, this book aspires to inspire further research, innovation, and collaborative efforts toward sustainable agricultural systems.

I express my sincere gratitude to Dr. Himanshu Pathak, Secretary (DARE) and Director General (ICAR), and Dr. S.K. Chaudhari, Deputy Director General (NRM), ICAR, New Delhi, for their invaluable support and for writing the foreword to this book. I also extend my appreciation to the Director, ICAR-Indian Institute of Soil Science, Bhopal, and all my colleagues for their cooperation throughout the preparation of this volume.

I am deeply grateful to the contributors for their dedication and expertise, which have enriched the content of this book. Additionally, I extend my thanks to the scientists and staff of the institute for their direct and indirect assistance in preparing the manuscript. It is my sincere hope that this publication will be a valuable resource for students, researchers, policymakers, planners, farmers, and environmentalists alike. I believe that *Nanofertilizers in Agriculture* will serve as a crucial reference, fostering the advancement of nanofertilizer science and application and empowering stakeholders to embrace innovations that shape a more sustainable future for modern agriculture.

Bhopal, Madhya Pradesh, India Ashok K. Patra

# Contents

# Series Editor and Editorial Board Members

## About the Series Editor

**Rajeev Kumar Varshney** FRS, is a renowned agricultural scientist with over two decades of experience in genomics and molecular breeding. Currently, he holds multiple prestigious positions at Murdoch University in Australia. He has made a significant impact on global food security by creating genomic resources for major "orphan" tropical crops, using these resources to identify genetic loci and candidate genes for drought and pest tolerance in staple crops in sub-Saharan Africa and India, and spearheading international programs that deliver superior crop varieties to some of the world's most impoverished farmers. He is an accomplished author with more than 600 publications to his name and has been recognized as a Clarivate Highly Cited Researcher for ten consecutive years (2014–2023).

Food Futures Institute, Centre for Crop and Food Innovation, State Agricultural Biotechnology Centre, Agriculture and Food Security, Murdoch University, Perth, WA, Australia.

## Editorial Board Members

## About the Editor

**Ashok K. Patra** is an eminent soil scientist with a distinguished career spanning over three decades. He earned his bachelor's degree in agriculture from Banaras Hindu University, India, followed by an M.Sc. and Ph.D. in soil science from the Indian Agricultural Research Institute (IARI), New Delhi. Since 1989, he has been associated with several premier institutions under the Indian Council of Agricultural Research (ICAR) and has collaborated with various international organizations. His professional engagements have taken him to multiple countries, contributing significantly to global soil research and policy discussions.

Dr. Patra is globally recognized for his significant contributions to soil organic carbon dynamics, nitrogen transformations, soil ecology and biodiversity, and nanofertilizers for sustainable soil management in diverse agro-ecosystems. His research has earned him numerous prestigious accolades, including the UN FAO Glinka World Soil Prize (2022), the ICAR Rafi Ahmed Kidwai Award (2021), the IARI Hooker Award, the IPNI-FAI Award, and the Platinum Jubilee Commemoration Award from the Indian Society of Soil Science (ISSS). He is a Fellow of the National Academy of Agricultural Sciences (NAAS), the National Academy of Sciences, India (NASI), and several other esteemed professional societies.

His global impact is further underscored by his recognition as one of the 13 most influential soil scientists worldwide (Food Tank, USA, 2017) and his ranking among the top 2% of scientists globally in agriculture, fisheries, and forestry (Stanford and Elsevier 2024). For his significant efforts in raising mass awareness about soil health and sustainable management, ICAR-IISS, Bhopal, under his leadership, was honored with the UN FAO King Bhumibol World Soil Day Award (2020).

Dr. Patra has also held key leadership positions, serving as the President of the Indian Society of Soil Science and the Indian Science Congress (Agriculture and Forestry Sciences Section). Additionally, he has contributed as the chairman/member of various national and international scientific committees. His prolific academic contributions include the publication and editing of 18 books, 260 research journal articles, with over 10,000 citations, an h-index of 48, an i-10 index of 192, and a total of more than 400 publications.

Before assuming the position of Vice Chancellor at Bidhan Chandra Krishi Viswavidyalaya (BCKV), Dr. Patra served as the Director of ICAR-Indian Institute of Soil Science (IISS), Bhopal (2014–2022), leading the institute to new heights. He later continued his scientific endeavors as an ICAR-Emeritus Scientist (2023–2024).

## Contributors

**Tapan Adhikari** ICAR-Indian Institute of Soil Science, Bhopal, Madhya Pradesh, India

**Suparna Biswas** Department of Physics, Indian Institute of Engineering Science and Technology, Howrah, West Bengal, India

**Vassanda Coumar M** ICAR-Indian Institute of Soil Science, Bhopal, Madhya Pradesh, India

**S. P. Datta** ICAR-Indian Institute of Soil Science, Bhopal, Madhya Pradesh, India

**Avijit Ghosh** ICAR-Indian Grassland and Fodder Research Institute, Jhansi, Uttar Pradesh, India

**Tejendra Krishna** Department of Soil Science and Agricultural Chemistry, Bihar Agricultural University, Bhagalpur, Bihar, India

**Dhiraj Kumar** ICAR-Indian Institute of Soil Science, Bhopal, Madhya Pradesh, India

**Jitendra Kumar** ICAR-Indian Institute of Soil Science, Bhopal, Madhya Pradesh, India

**Manoj Kumar** ICAR-National Research Centre for Makhana, Darbhanga, Bihar, India

**Nintu Mandal** Department of Soil Science and Agricultural Chemistry, Bihar Agricultural University, Bhagalpur, Bihar, India

**Rahul Mishra** ICAR-Indian Institute of Soil Science, Bhopal, Madhya Pradesh, India

**Sukanya Misra** Rani Lakshmi Bai Central Agricultural University, Jhansi, Uttar Pradesh, India

**Ashok K. Patra** ICAR - Indian Institute of Soil Science, Bhopal, Madhya Pradesh, India

**D. Pradeep** Center for Agricultural Nanotechnology, Tamil Nadu Agricultural University, Coimbatore, Tamil Nadu, India

**M. Praghadeesh** Centre for Agricultural Nanotechnology, Tamil Nadu Agricultural University, Coimbatore, Tamil Nadu, India

**S. K. Rajkishore** Department of Environmental Sciences, Tamil Nadu Agricultural University, Coimbatore, Tamil Nadu, India

**Binoy Sarkar** Future Industries Institute, University of South Australia, Adelaide, Australia

**Satdev** Department of Soil Science and Agricultural Chemistry, Bihar Agricultural University, Bhagalpur, Bihar, India

**Amit K. Singh** ICAR-Indian Grassland and Fodder Research Institute, Jhansi, Uttar Pradesh, India

**Anand Singh** Acharya Narendra Deva University of Agriculture and Technology, Ayodhya, Uttar Pradesh, India

**Nishant Kumar Sinha** ICAR-Indian Institute of Soil Science, Bhopal, Madhya Pradesh, India

**G. Srinivasan** Center for Agricultural Nanotechnology, Tamil Nadu Agricultural University, Coimbatore, Tamil Nadu, India

**K. S. Subramanian** Centre for Agricultural Nanotechnology, Tamil Nadu Agricultural University, Coimbatore, Tamil Nadu, India

Coromandel Nanotechnology Center, Coimbatore, Tamil Nadu, India

**J. C. Tarafdar** ICAR-Central Arid Zone Research Institute, Jodhpur, Rajasthan, India

# Chapter 1
# Nanofertilizers: An Overview

**J. C. Tarafdar**

**Abstract** Global hunger index is increasing day by day due to population outburst and marginal increase in food production as well as running out of fertile agricultural land. Therefore, immediate attention of increase in productivity through innovative technology is needed as the land and water resources are limited gradually. Among all the technology available at present nanofertilizer technology is the most potential one for sustainable improvement in agricultural sector. Nanoparticles ($10^{-9}$ m) have more surface area to volume ratio and have more penetration and triggering potential which make them very much applicable in agriculture. It has the potential to use as nanofertilizer for crop production which can release the nutrients slowly as per crop demand. Nanofertilizers can be prepared through physical, chemical, aerosol, and biological techniques. It can increase the nutrient use efficiency, mobilization of native nutrients, soil aggregation, water retention, site targeted delivery, reduction of toxicity, plant abiotic and biotic stress management, and enhanced nutrient utilization. Nanofertilizer may be used as foliar application, in the soil, seed soaking, through drip, sprinkler, hydroponic, aeroponic, and aquaponic. Nanofertilizer requirement is very low as compared to the chemical and organic fertilizers and have the potential to build up soil carbon. Besides elevation of antioxidant enzyme activities, they help in plant tolerance and defense mechanism. It plays a role in improving the photosynthetic pigments and reducing $H_2O_2$ and MDA levels. Nanofertilizer has definite role to alter plant metabolic activities and help in improving membrane permeability and stability. Application of nanofertilizers encourages higher seed germination, plant vigor index, plant biomass, chlorophyll level, hormonal control, photosynthetic rate as well as yield attributes. It is more cost-effective than traditional fertilizers as their requirement and frequency of application is low as compared to them. It has the potential to play a major role in sustainable agriculture and to meet the global food demand. But there is a need for more research on ethical issues of nanofertilizers in long-term basis for the ultimate effect on soil, human, and environmental health.

---

J. C. Tarafdar (✉)
ICAR-Central Arid Zone Research Institute, Jodhpur, Rajasthan, India

 
A. K. Patra (ed.), *Nanofertilizers in Agriculture*, Sustainability Sciences in Asia and Africa, https://doi.org/10.1007/978-981-97-8112-6_1

**Keywords** Nanofertilizers · Types · Synthesis · Application · Stress management · Health management · Safety issues

## 1.1 Introduction

To reduce the global hunger index, there is an urgent need for transformation in agricultural system as well as effective inputs as early as possible. Nanofertilizer technology may play a major role to support agricultural production. Nanoparticles may play versatile roles that can be widely used as nanofertilizer or nanonutrients for agricultural purpose. It has the potential to improve the soil fertility for better yield, nutrient use efficiency besides increasing crop quality and productivity. Nanomaterial may enable controlled release ensuing slow diffusion into the soil. It may contain single or more macro- or micronutrients. The major advantages of nanosize particles are they have large surface area to volume ratio resulted ideal for holding and releasing nutrients. It has the potential to reduce runoff and nutrient leaching in the environment. They can penetrate better into the cell and help to increase plant and microbial activities that influence the nutrient use efficiency. They are the effective catalyst of plant and microbial metabolism and ability to trigger the enzyme release. Based on their action and nutrient composition nanofertilizers may be classified as targeted delivery nanofertilizers, controlled release nanofertilizers, plant growth stimulating nanofertilizers, nutrient loss preventing nanofertilizers, hybrid nanofertilizers, nutrient loaded nanofertilizers, inorganic and organic nanofertilizers, consistency-based nanofertilizers as well as carrier-based nanofertilizers. This chapter aims to express the scope, type, synthesis, and application of nanofertilizers, their response on management of soil and plants as well as the safety issues and their environmental concern to sustainable use in agriculture.

## 1.2 Scope of Nanofertilizers

Nanofertilizers have enormous scope to use in agriculture. It may be used as nutrients, pesticides, sensors, coating material, catalyst, barcodes, filters, tracers, carriers, and so on. The various scope of nanofertilizer use in agriculture is presented in Fig. 1.1.

Nanofertilizer can maintain balanced nutrient supply, stimulated crop growth as well as quality improvement. It has higher efficiency, biodegradability, better adoption to drought and salt stress, capability to ROS accumulation, and help in stress gene expression. They need low rate of fertilizer application with more nutrient use efficiency. They can extend the duration of nutrient supply into the soil for plant growth. Nanosize micronutrient formulation may help to improve solubility and dispersion of insoluble nutrients in the soil resulted less absorption and fixation thereby increasing the bioavailability. They help to mobilize 30% more native

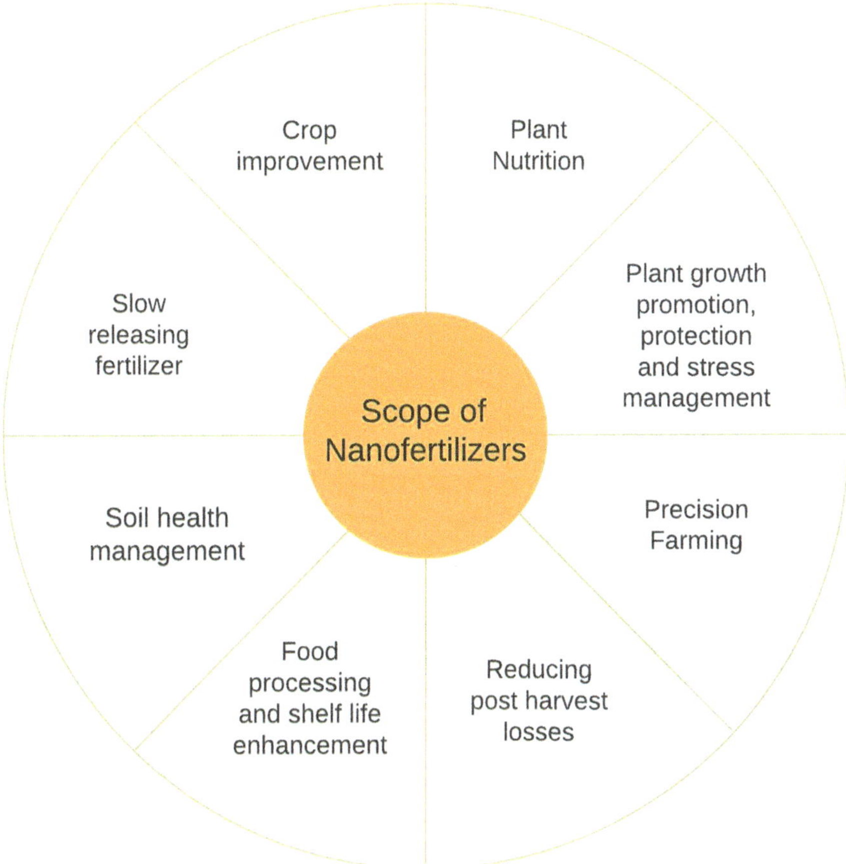


**Fig. 1.1** Scope of nanofertilizer use in agriculture

nutrients for plant production (Tarafdar 2021). They have negligible nutrient loss with 3–20 times more efficiency of different nutrients. Nanofertilizer can effectively use as seed coating and pesticide carrier. They are the effective material for precision farming and nanobioformulations. It can be used as soil conservation (such as nanozeolites) and enzyme activation for plant and microbes. Nanosize particle-induced microbial polysaccharides are very effective for soil aggregation, moisture retention, and soil carbon build up even under harsh soil conditions (Raliya et al. 2014). Nanosensors, nanopolymers, nanoclays, and nanozeolites may be used for contaminant detection in soil and water. Nanomembranes can be used for water purification, desalinization, and detoxification. Nanomolecules are useful to prevent water leakage in robust water tank. Porous nanomaterials may be used to minimize nitrogen loss after balancing demand-based release.

## 1.3 Types of Nanofertilizers

Nanofertilizers may be carbon based, metal based, or polymer based. Again it may be categorized as controlled release, targeted delivery, action based, water and nutrient loss controlling, plant growth stimulating, and soil health management. Controlled release nanofertilizers can very well overcome the problem of conventional fertilizers like insufficient nutrient use and nutrient leaching (Liu and Lal 2015). They are capable of improving nutrient uptake as well as reduce environmental stress (DeRosa et al. 2010). It may capture nutrients in nanoscale carrier materials like polymers, lipids, or inorganic substances. The release of nutrients from nanofertilizers may be influenced by many factors including pH, temperature, soil moisture, and enzyme levels. It is an effective material to reduce contamination of water bodies and overcome eutrophication to protect aquatic ecosystems (Babu et al. 2022). They are also familiar to improve in crop yield and nutrient use efficiency and good material to meet global demand. Carbon-based nanosize materials easily absorb micronutrients from the soil for the release in future that way they may help in seed germination and plant growth after keeping soil remain wet and nourished (Saxena et al. 2014). It may improve the nutrient delivery as well as uptake which ultimately help in overall growth of the plants. The most common use of carbon-based fertilizer is carbon nanotubes which found to accelerate the nutrient delivery and uptake resulted more crop productivity (Khodakovskaya et al. 2009). Besides this, people are using graphene as a carbon-based nanofertilizer that is found to be effective in nutrient use efficiency and plant growth (Zhang et al. 2022). Chitosan-based nanofertilizers may serve also a controlled release fertilizer for agricultural use. It may also be used as plant growth regulators, as well as herbicides and insecticides (Mujtaba et al. 2020). Chitosan as a carrier help to protect biomolecules from extreme light, temperature, and pH as well as shielding plant cells from burst release. It has potential to form complexes with metal ions resulted reducing their toxicity. They can be used as a slow-release nanofertilizer. Potassium incorporated chitosan nanoparticles can effectively be used to prevent soil degradation and nutrient losses (Kubavat et al. 2020). It may raise soil porosity, water conductivity, and soil friability. They may provide soil coating to reduce degradation and increasing productivity after synchronizing crop demands and reducing fertilizer requirements. Different nanofertilizers have dissimilar properties resulted varies to the plant effects. They can be classified easily depending on the carrier nutrients, their performance and consistency. The various types of nanofertilizers are presented in Fig. 1.2.

Clay-based nanoparticles have nanolayer reactivity and may be utilized as controlled release fertilizer formation. Their active surface can be utilized in physicochemical and biological processes (Sempeho et al. 2014). Nanofertilizer may be prepared as nanocapsules which are composed of inorganic and organic materials from lipids, biopolymers, silica, metal oxides, and carbon nanotubes. The advantages are they may form a protective barrier and act as a slow-release fertilizer. They have more efficient fertilizer use and greater control with stability with reduce

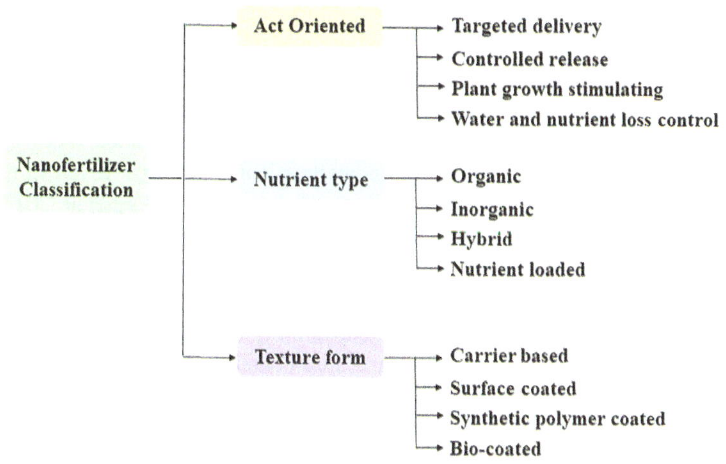


**Fig. 1.2** Category of nanofertilizers use in agriculture

leaching risk. Nanogels can be infused with indispensible elements like N, P, and K for improving plant growth (Krishnani et al. 2022). Besides fertilizers, it may be used as water purifiers and drug delivery. Calcium phosphate nanogels are reported to improve seed germination rate, as well as increase in beneficial enzyme activities like amylase, protease, and nitrate reductase (Umarani and Mala 2013). The polyurethane-based nanofertilizers which are produced by metathesis polymerization are reported to be effective in delivering different nutrients to the soil (Thirugnanasambandan 2021). They may shield the particles from the environment and release in a controlled manner. Besides this, it may help to increase the water holding capacity and reduce soil erosion. It may be very useful for future application in agriculture. Starch-based nanofertilizers are mainly nanocrystals extracted from starch and can be applied to plants both in liquid and aerosol form (Lin et al. 2011). They are found very effective to grab reactive oxygen species. Zeolite-based nanofertilizers have the capacity to retain water, nutrients, and organic compounds. It may penetrate deeper into the soil and can efficiently supply the nutrients to the root zone (Sharma et al. 2022). Moreover, zeolites may operate as reservoirs for storing and releasing nutrients as and when the plant needs and ultimately help to reduce the cost of fertilizer application. Chitosan-based nanofertilizers have the advantage for controlled release and can merge with other nutrients and agrochemicals that make them a good candidate for sustainable agriculture.

Nanosensor can be effectively used as precision farming, pesticide detection, and evaluation of plant pathogen (Kaushal and Wani 2017). It may supply fertilizer to plants after attaching fertilizer component to a nanoparticle. It also helps the nanoparticle to enter the plant cell and can precisely measure the required amount of fertilizer. But detail information is needed before their commercial application. Besides this, nano-coated urea, nano hydroxyl apatite, nano-nutrient carriers, nanoemulsions, and clay-based nanofertilizers are found to be very effective (Liu and

Lal 2015). It may help to control release rate as well as get rid of environmental impacts (Kottegoda et al. 2011). Some of the nanofertilizers like carbon nanotubes can interact with plant root system to stimulate plant growth after increasing hormone synthesis. They help to increase both carbon and nutrient content in the soil after absorbing and releasing the nutrients to the appropriate place. They also help to build soil structure; water retention as well as water transport result in reducing soil erosion and have the ability to interact with plant cells (Aslani et al. 2014). They are good material for seed coating and help to overcome the pest attack and assist the seed to absorb more water and nutrients. They are good material to resist the heavy load and aid in improving crop yield. Some of the nanofertilizers can help to improve water and nutrient loss after encapsulating or modifying the surface to be hydrophilic. The important examples are urea coated with iron oxide, sulfur, calcium, magnesium, zinc, etc. nanoparticles (Ahmed and Fahmy 2017; Jakhar et al. 2022). Two useful slow-release nanofertilizers are in the form of nanobeads and nano-emulsions. Nanobeads can be made after using iron, carbon, or other metals. They can make very tiny cracks in the soil and assist to fertilize the plants. They can also help to clean up pollution. NanoFert and N-Flex are the commercially available nanobeads material. Nano-emulsion-based fertilizers are generally manufactured after combining surfactant with liquid to produce a stable homogenous mixture. Due to minute droplet size, they help to enhance plant cell walls permeability and rapid fertilizer absorption (Thiruvengadam et al. 2018) and can be adjusted to supply specific nutrients to plants. They are reported to overcome fungal pathogens and economic benefit to the crops (Krishnamoorthy et al. 2021). They can also be used to deliver rhizobacteria.

Nutrient-based nanofertilizers are able to increase nutrient availability, uptake as well as utilization in plants. They may be metals, metalloids, or non-metallic nanoparticles. Industries are now producing and marketing these products. They may be macronutrient or micronutrient-based fertilizers. Inorganic nanofertilizers may reduce eutrophication as well as emission of greenhouse gases (Mejias et al. 2021). Potassium-based nanofertilizers can improve physical properties of soil besides the increase in absorption rate and resistant to leaching. They have tremendous absorption and penetration rate and very effective in achieving higher yields (Sheoran et al. 2021a, 2021b). Calcium-based nanofertilizers are useful to resist diseases and improve the quality of fruits and vegetables (Shilpa et al. 2022). In general, the effect of nanomaterials depends on the concentration used. They may also influence molecular mechanisms and signaling pathways. It has been reported that NPK nanofertilizers are very effective in precision agriculture and increase in fertilizer use efficiency (Ramirez-Rodriguez et al. 2020). Important magnesium-based nanofertilizers are magnesium oxide and magnesium sulfate that are reported to enhance the quality of fruits and vegetables and found to be effective against diseases and pests (Liao et al. 2021). Chitosan along with magnesium nanofertilizers may overcome drought stress (KhordadiVaramin et al. 2018). Many types of sulfur-based nanofertilizers have been developed. They are found to be effective as slow-release fertilizers besides playing an important role in plant defense mechanisms protecting diseases and pests. Micronutrient-based nanofertilizers like B, Cu,

Zn, Ni, Ti, Mo, etc. play a useful role in plant health, growth, and development. They can also be used to overcome plant biotic and abiotic stresses. Nano Zn chelate as well as nano B chelate fertilizers are also found to be very effective (Davarpanah et al. 2016). Cu nanofertilizers are equally effective in resisting pests and diseases besides boosting crop yield (Chhipa and Joshi 2016). Iron-based nanofertilizers may help to regulate the release of nutrients in a way that is coordinated with nutrient demand of the crops as well as improve water quality under aquatic environments. Ni-based nanofertilizers are found to help plant growth and development. Ti-based nanoparticles may be used as an antimicrobial agent and more sunlight harvest. Zn-based nanofertilizers may accelerate plant metabolic activities and help to reduce the environmental contamination (Sheoran et al. 2021a, 2021b). Nanoparticles which are derived from organic materials may be termed as organic nanofertilizers. They are mainly originated from the natural sources. They may be capsules, polymer, micelles, vesicles, liposomes, conjugates, polymersomes, dendrimers, and polymeric nanoparticles (Fatima et al. 2021). Organic nanofertilizers may work as slow-release fertilizers which are very effective in plant absorption and use. It can also improve soil health by improving soil aggregation, porosity, water infiltration, aeration, and root penetration (Lehmann and Kleber 2015). It can also be made from plant waste and compost. NanoMax-NPK is a good commercially available organic nanofertilizer. Hybrid nanofertilizer is made up of a combination of conventional and nanotechnology-based fertilizers. They can be used as a slow-release fertilizer which improve efficiency and reduce environmental impact (Tarafder et al. 2020). Nanocomposites are prepared by combining a polymeric matrix and inorganic nanomaterial. They have more strength, toughness, and rigidity besides higher storage stability and pH tolerance (Aththanayaka et al. 2022). They are very good material for nano fertigation and can be applied directly to the plant root system resulted more nutrient uptake and reduced waste. The most important hybrid nanofertilizers are nano-encapsulated and nano-clay fertilizers.

The important nutrient-loaded fertilizers are nano-porous zeolites. It can work both as fertilizers and soil amendments. It has more water retention capability which results in reduced irrigation for optimum crop production. The consistency-based nanofertilizers are manufactured using surface-coated and carrier-based technologies. They may be synthetic polymer-coated or biological product-coated nanoparticles (Vejan et al. 2021). Mesoporous silica nanoparticle is a very good example of carrier-based nanofertilizers that can enhance nutrient solubility, availability, and stability resulting in an increase of crop yield (Kong et al. 2021). Nanoparticles like silica and iron are good materials to prepare surface-coated nanofertilizers. Gold, silver, carbon, and titanium dioxide can also be used for surface-coated nanomaterials. They have more efficiency to their utilization and results improved plant growth. Synthetic polymer-coated nanofertilizers are coated with a thin polymer which can protect nanofertilizers from degradation and better handling. They can also prevent nutrient loss. The important commercial products are Nutricote, Osmocote, and Polyon (Singh et al. 2021). Polyvinyl-coated zeolite is also very popular under this group. Biological coated nanofertilizers are more benefited in agriculture as they improve soil water retention and microbial activity in many folds. Clay-based nanofertilizers are more popular in the market due to their simplicity of application and even coverage on plant surfaces (Wilson et al. 2008). Microbial coated nanofertilizers or nanobioformulations efficiently transport the nutrients to plant roots and tissues. They may also help in nitrogen fixation, phosphate solubilization as well as replenishment of soil nutrients (Sharma et al. 2023). Nanobioformulations are found to be very useful materials for cereals, legumes as well as horticultural crops (Kalia and Kaur 2019). Carrier-based nanofertilizers can improve nutrient availability and reduce environmental impact. The important carrier now use as nanofibers, polymer-nanocellulose-clay composites, carboxy methyl cellulose, and silk-fibroin-derived nanocarriers.

## 1.4 Synthesis of Nanofertilizer

Nanofertilizers may be synthesized singly or in combinations. They can be synthesized by physical, chemical, aerosol, and biosynthesized technique. There may be top-down and bottom-up approach. The synthesized nanoparticles can be of different shapes depending on the methods used and the sources they originate from. The important physical methods used for preparation of nanofertilizer are grinding, thermolysis, sputtering, pulsed laser deposition technique, condensation method, and microwave-assisted synthesis technique. Natural deposited minerals are used as nanofertilizers after purification and grinded to nanosize ($<10^{-9}$ m) by ball mill or pot mill. In general, more than 350 °C is required for production of nanofertilizer by thermolysis process. Sputtering is an electron-beam lithography system; it is widely used for manufacturing of metal nanoparticles. Pulsed laser deposition technique is widely accepted by the industries for the production of ZnO, $TiO_2$, and boron nanoparticles. Condensation method is used in various ways for the preparation of nanofertilizers such as metal evaporation, laser vaporization, thermal vaporization, plasma vaporization, and solar energy-induced vaporization. Microwave-assisted synthesis is basically an energy conversion method rather than heat transfer to the system where heat energy converts electromagnetic energy into thermal energy.

There are many chemical methods to prepare nanofertilizers. The widely used chemical methods are sol gel method, poly vinyl pyrolidene (PVP) method, co-precipitation technique, micro-encapsulation method, sonochemistry, colloidal method, hydrothermal synthesis, and micro-emulsions method. Sol gel technique involves chemical transformation of liquid (sol) into a gel state. It is actually a wet chemical technique and takes place in three stages like hydrolysis, condensation and polymerization, particles growth. The entire process may be divided into stages like hydrolysis, polymerization, growth of nanostructures, and agglomeration. PVP method is a solution phase synthesis of nanofertilizers of Ag, Au, Pd, etc. The morphology of the particle may be adjusted here by changing molar ratio of PVP to the salt precursor. Co-precipitation technique is another important chemical synthesis technique where reactions entail the concurrent occurrence of nucleation, growth, coarsening, and agglomeration process. There may be microwave or sonication-assisted co-precipitation. In micro-encapsulation technique, nanoparticles may be encapsulated inside a micron-sized matrix. It is a popular method for industries to produce $TiO_2$ nanoparticles. Under sonochemistry high frequency sound wave is used to prepare nanofertilizers. Colloidal method is very popular in preparing nanofertilizers. Different shapes (cubic, tetrahedral, spherical, octahedral, etc.) of nanofertilizers can be prepared by this method. Hydrothermal synthesis has been performed in the sealed vessel like bomb, autoclave, etc. Autogenous pressures are used to bring solvent temperature well above their boiling points. Micro-emulsion methods are more popular for preparation of magnetic nanoparticles to be used as nanofertilizers. It is driven by diffusion-based process.

Aerosol method of nanofertilizer synthesis is very attractive due to their more product purity and potential of scale up. Here, by changing carrier gas flow as well as heater and diffusion drier sizes, different shapes and sizes of nanofertilizer particles can be produced. The important aerosol techniques are furnace method, flame method, electrospray technique, chemical vapor deposition (CVD) technique, and physical vapor deposition (PVD) method. Generally, tube furnace (500–1000 °C) is used for nanofertilizer synthesis. Flame method of nanofertilizer production has been recognized as the most commercially successful method which produces millions of metric tons of $TiO_2$, carbon black, and other metal oxide nanoparticles. Electrospray method is very accurate in producing desired shapes and sizes of particles. The equipment consists of an atomizer, carrier gas, reactor, and sampler. It is used well in food industry. Under CVD process, the vaporized precursors are introduced into a CVD reactor at elevated temperature. It is very useful for synthesis of carbon nanomaterials (CNMs) in terms of large-scale production and maintains its purity. The PVD technique entails the condensation from the vapor phase. The important processes involved are thermal evaporation and sputtering.

Fungi, bacteria, actinomycetes, algae, yeast, and plant extracts can synthesize nanofertilizer particles biologically in environment-friendly manner. The possible mechanisms of nanofertilizer production by different agents are presented as Table 1.1.

Fungi are well recognized for synthesizing all types of fertilizer particles in nanosize needed by plants. Generally, 32 kDa protein released from fungi is believed

**Table 1.1** Mechanisms of biological nanofertilizer production

| Representative | Mechanism |
|---|---|
| Fungi | By fungal protein produced both intra- and extracellularly and the procedure of biomimetic mineralization |
| Bacteria | After reducing metal ions through the specific reducing enzymes like NADH-dependent reductase or nitrate-dependent reductase |
| Actinomycetes | Reduction and stabilization through enzymes, organic acids, and bioactive compounds |
| Algae | Reduction and stabilization through hydroxyl groups and other functionalities |
| Yeast | Membrane bound and cytosilicoxido-reductases and quinine |
| Plant | Alkaloids, flavonoids, saponins, tannins, etc. act as reducing and stabilizing agents |

to be responsible for nanofertilizer synthesis. It has been noticed that some NADH-dependent reductase also involve in reduction of some of the nanofertilizer particles. A large number of fungi have been reported to be capable of producing N, P, K, Ca, Mg, S, B, Cu, Fe, Mn, Mo, and Zn from their respective salt solutions. Generally, metal nanoparticle fertilizers are synthesized by bacteria with the help of NADH-dependent reductase. Actinomycetes are very efficient in manufacturing polydispersed, stable, and ultra-small size of nanofertilizer particles through the production of enzymes, organic acids, and bioactive compounds. They are very useful in synthesizing metal nanoparticles like Ag, Au, Zn, Cu, and selenium. Algae are well known for synthesizing nanoparticles like palladium, gold, iron, and silver. Yeast is well known for synthesizing zinc phosphate and cadmium sulfide nanoparticles. They may be used for synthesis of other metal nanoparticles with the help of their reducing enzymes produced intracellularly or extracellularly. Plant extract (leaf, stems, roots, fruits) is found to be useful for synthesis of Ag, Pt, Au, Pd, Pb, Zn, Cu, and Fe nanoparticles. Biological and phytonanotechnological synthesis of nanofertilizer particles are ecofriendly, simple, rapid, stable, and cost-effective.

## 1.5 Application of Nanofertilizer

Nanofertilizer can be applied as foliage, in the soil, through drip, sprinkler as well as in vivo methods like hydroponic, aeroponic, and aquaponic besides seed treatment. But they are found to be more efficient when applied as foliage due to the triggering effect of beneficial plant enzymes to mobilize more native nutrients for plant nutrition. When applied as foliage, the nanofertilizer particles may enter through the cuticle or epidermis, stomata, hydathodes, stigma, or other entry sites of plant shoots. The nanofertilizer particle may also enter through roots such as root tips, rhizodermis, lateral root junctions, cortex, or any wounding under hydroponics or soil application. The movement of nanofertilizer particles inside the plant cells also depends on their concentration, size, and shapes (Tarafdar et al. 2012). For controlled release of nanofertilizer, polymeric nanoparticles are the best. The important polymers are polylactide, polycaprolactones, and polyacrylates. Besides this, natural polymers like chitosan, alginate, albumin, etc. can also be used. Soil applications of nanofertilizers are mostly farmers' choice, but it has limitations in respect of native nutrient mobilization and nutrient use efficiency. Drip application of nanofertilizer particles is more effective than soil application. It is better to use micronutrients in nano form under drip because most of the micronutrients are not readily translocated within the plants. Moreover, this process of feeding is found to be very efficient when the roots cannot provide the required nutrients. Drip application also helps to minimize adverse environmental impact. Nanofertilizer can be applied as sprinkler with reduced concentration as under sprinkler the environmental losses are more. It is only suited for row field and tree crops. Nanofertilizer application to hydroponic is needed to calculate the concentration of nanoparticles in the solution as well as demand of oxygen. The pH as demand should maintain throughout the plant growth period. Nanofertilizer use under aeroponic found to be very effective for optimizing nutrient utilization with minimum losses. It is found to be very effective for horticultural crops (Sharma et al. 2018). Nanofertilizer may be applied under aquaponic to grow fish under closed environment. It may have large commercial application. Nanofertilizers like Cu, Ag, Fe, $TiO_2$, Ni, etc. is found to be a very effective material for seed treatment. It may promote the seed germination and plant growth besides protecting seeds from soil-borne pathogens. It can also apply as dusting, fogging, and emulsion.

## 1.6 Response of Nanofertilizer

Nanofertilizer particles may enter into the plants through various opening such as cuticle, epidermis, stomata, hydathodes, stigma, root tips, rhizodermis, lateral root junction, and wounding. They may move through the cell sap after entering and while moving they can trigger the different plant co-enzyme systems and help plants to release more beneficial enzymes for native nutrient mobilization and ultimately plant nutrition. They may help plants to improve nutrient uptake, water holding capacity, soil quality, carbon sequestration, microbial activity, plant productivity and health, environmental quality and abiotic as well as biotic stress management (Tarafdar 2021). They also act to reduce nutrient volatilization, increase nutrient use efficiency, and reduce nutrient leaching. They can be used as slow and controlled release fertilizer and bionanoformulations. In general, their requirements is 80–100 times less than chemical fertilizer due to their 4–20 times higher nutrient use efficiency and 30% more native nutrient mobilization. It has been noticed that with the application of nanofertilizer the plant becomes 10 times more stress tolerant than chemical fertilizer or no fertilizer application. They help in seed germination and protect plants against oxidative stress as they can mimic the role of antioxidative enzymes like superoxide dismutase (SOD), catalase (CAT), and peroxidase (POX). They may help to increase plant metabolites, enzyme activities, and antimicrobial properties of plants which resulted positive effect on plant growth and development. Table 1.2 shows their overall effect on dry matter and yield with some of the selected crops.

Nanofertilizer may be responsible for different morphological and physiological changes of plants which depend on the type, nature, and doses of the nanofertilizer use. It may help plant photosynthetic activity and enhances electron transport rates. It may help to increase root activity and significant effect on chlorophyll a and b as well as carotenoid content (Tarafdar 2021) while decreasing $H_2O_2$ and MDA content.

**Table 1.2** Effect on dry matter and grain yield of some of the selected crops over recommended doses of chemical fertilizer (RDF)

| Crop | Nanofertilizer use | % increase in dry matter over RDF | % increase in grain yield over RDF |
|---|---|---|---|
| Clusterbean | P, Zn, Fe, Mg | 19–33 | 17–37 |
| Maize | P, Zn | 18–31 | 23–38 |
| Moth bean | P, Zn, Fe | 18–30 | 17–31 |
| Mung bean | P, Zn, Fe, Mg | 20–39 | 17–42 |
| Pearl millet | P, Zn, Fe | 22–40 | 18–48 |
| Tomato | P, Zn | 16–25 | 18–29 |
| Wheat | P, K, Zn | 21–38 | 22–32 |

## 1.7 Nanofertilizer for Plant Stress Management

Large numbers of nanofertilizer particles are found to be very effective in controlling both biotic and abiotic stresses. Abiotic stresses like drought can be controlled by $TiO_2$, ZnO, Fe, $SiO_2$, Ag, Analcite, and Maghemite nanofertilizer particles (Tarafdar 2021). They may help to increase germination percentage, activation of CAT, accumulation of flavonoids and carotenoids, photosynthetic pigments and biomass accumulation, as well as chlorophyll, growth and yield attributes besides decrease in proline accumulation, $H_2O_2$ and MDA content. Nanofertilizer particles may also activate stomatal opening, increased $CO_2$ assimilation, and mobilization of trace elements. They help to increase plant growth and yield besides improvement in gluten and starch content. Nanofertilizer particles like $SiO_2$, ZnO, and $Fe_3O_4$ may help to overcome salinity stress suffered by plants by increasing seed germination, root length, leaf water content, and seedling growth. They help to reduce chlorophyll degradation and oxidative damage and enhanced photosynthetic parameters as well as antioxidative enzymes. Abiotic stresses like cold may overcome with the application of Se, $TiO_2$, $SiO_2$, and Ag nanofertilizers which are helping to modify the physiological response of plants, increase proline content in leaves, reduce lipid peroxidation in roots, enhance activities of antioxidant enzymes, decrease $H_2O_2$ content and electrolyte leakage, and enhance activity of phosphoenolpyruvate carboxylase. Heat and temperature stresses may be overcome with the application of $TiO_2$, MWCNTs, and $Na_2SeO_4$ nanofertilizers. They help in up-regulation of the expression of various stress-related genes. They may help to enhanced photosynthesis after regulating energy dissipation resulted cooling of leaves through inducing stomatal opening. They also help to enhance plant growth, chlorophyll, and leaf relative water content. $TiO_2$ nanofertilizer application also found to reduce heavy metal like Cd toxicity to plants. Besides this, Zn, Si, $TiO_2$, and Fe nanofertilizers are found to be effective in overcoming different heavy metal stresses. Nanofertilizers like Zn and Fe are deemed to be very effective in overcoming moisture stress suffered by plants after influencing polysaccharide-producing microorganisms for the release of more microbial polysaccharide in the soil system which resulted in 10–14% higher moisture retention even under harsh arid environment (Raliya et al.

2014; Tarafdar 2022). Plants that received $CeO_2$ and $Al_2O_3$ nanofertilizers can overcome the flood stress. It was also found that Si and Ag nanofertilizer particles are very effective against abiotic stresses due to UV-rays.

The biotic stress may affect plant phytohormones and overall plant defense. Nanofertilizer particles have shown excellent antimicrobial properties to overcome plant biotic stresses. To get rid of insect stress, nanofertilizer-induced plants can go for cell wall modifications and activate the plant immune system. Normally, chitosan, Ag, MgO, $TiO_2$, and Si nanofertilizers are effective against different pathogenic attacks on plants. They can regulate the proteins provoking defense-related responses that in turn implicated in facilitating the activity of NADPH-oxidase which is the important ROS-producing enzyme during infections by pathogens (Satti et al. 2022). The biotic insect stress may be controlled with the application of nanofertilizers like CuO, chitosan with peppermint oil, ZnO, Ag, $SiO_2$, $Al_2O_3$, and $TiO_2$. They are mainly binding to S and P in proteins and nucleic acids that resulted in a decrease in membrane permeability that helps in organelle and enzyme denaturation and ultimately cell death. Similarly, herbivore attack may be controlled after $TiO_2$, Ag, MWCNTs, and Si nanofertilizer application. Application of these nanofertilizers may encourage plant resistance against herbivores after concurrently activating both plant physical and chemical defenses. Normally, nanoparticles help to enhance herbivore-induced plant volatiles (HIPV) production which activates the plant defenses against herbivores. The nanofertilizer application decreases herbivore attack and consumption rate either directly or indirectly. Mostly, it may act as a repellent to herbivores.

## 1.8 Nanofertilizer for Soil Management

Nanofertilizers have more surface area as well as fitting sorption properties; therefore, their chances of losses due to runoff is significantly reduced. Due to less phytodegradation, the uptake of nutrients into leaves and other parts of the plants is more. The particles are more stable as they possess low surface energy. Their aggregate size in solution mainly rests on the initial particle size and concentration. Nanofertilizer may help to reduce both nutrient leaching and nutrient volatilization. They may help to improve microbial activity and diversity in the soil. They may minimize soil toxicity as well as harmful effect of agrochemicals and chemical fertilizer application. With the continuous application of nanofertilizers through foliage, the carbon sequestration potential of the soil may improve; that way it may help to reduce environmental $CO_2$ concentration as well as soil carbon build up. Soil organic matter is found to stabilize nanofertilizer particles. Normally, the dissolve organic matter present to the surface of nanofertilizer particles can modify their surface physicochemical properties resulted increase in their stability. The soil carbon sequestration potential has been improved by 73% with the continuous application of nanofertilizers for 5 years. Figure 1.3 represents the effect on carbon

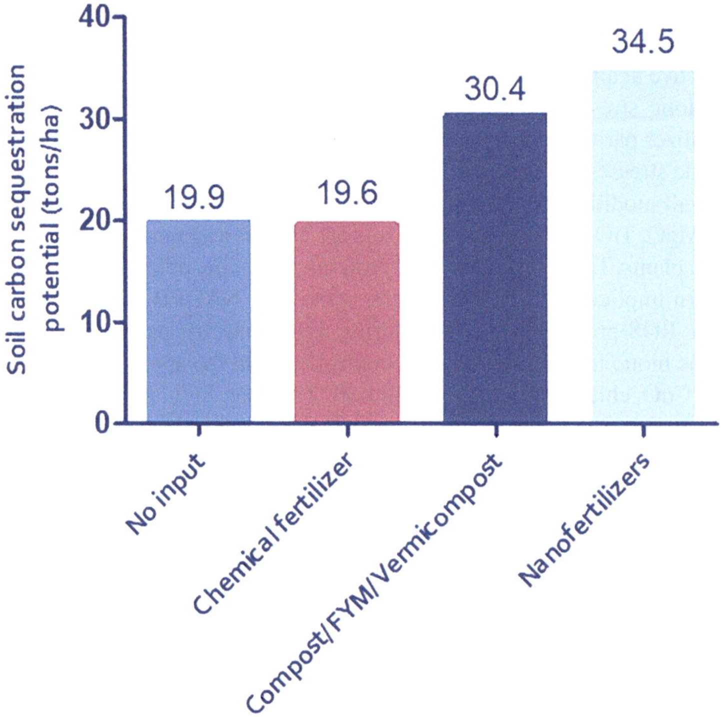


**Fig. 1.3** Soil carbon sequestrations potential after 5 years of continuous application under different treatments in alluvial soils

sequestration potential after continuous application of nanofertilizer for 5 years as compared to the chemical fertilizer, organic fertilizer, and no fertilizer treatment.

However, there are negligible changes with the soil organic carbon content (0.93–3.85%) with the application of nanofertilizer in soil. It has been reported that application of Au, $TiO_2$, Cu, Ag, Fe, Co, Ni, and carbon nanofertilizer particles may change microbial diversity and enzyme release (Tarafdar 2021). The major advantages of nanofertilizer over chemical fertilizers are they have more surface area, higher solubility, and more absorption capacity as well as penetration power. They have controlled release potential and more nutrient use efficiency. Their nutrient release potential also for longer period resulted in less frequent application of plants. They are effective in improving microbial activity in the soil. They help to increase soil activity after providing efficient delivery of nutrients to the root zone. They are also effectively increasing the activity of beneficial organisms which results in enhanced decomposition of organic materials and increased nutrient cycling (Tang et al. 2021). The improvement of soil activity by nanofertilizers helps in maintaining the soil ecosystem. Nanofertilizer also helps in improving soil structure and water holding capacity that create better soil environment for buildup of beneficial microorganisms. They can bind soil particles together forming larger aggregates and reducing soil erosion (Raliya et al. 2018). They have ecofriendly nature and low productive cost. They are the ideal material for precision farming improving plant stress tolerance and plant growth. They can penetrate the soil surface more effectively which resulted in better nutrient absorption and are more effective in stimulating plant growth.

## 1.9 Safety Issues of Nanofertilizer

Despite the huge potential of nanofertilizers, there are many unknown risks of nanofertilizers involving the environmental and health related. The risk associated with the global application of nanofertilizers is not fully evaluated. Nanofertilizers are most effective after foliage application but it needs a huge usable leaf area, and there is always the risk of scorching and burning if the concentration is high and recommended doses are not followed (Achari and Kowshik 2018). There is also need to apply at the right time. It is still not fully known whether nanofertilizers are converted into ionic form in the plants and then assimilate into proteins and other metabolites or some of them may reach through the food chain. The detailed study on this aspect is needed before global adoption. Very few studies so far conducted about the potential impact of nanofertilizer on human health as it has been observed that due to small size, organisms absorb nanoparticles very easily leading to cause damage to the gastrointestinal tract, liver, and kidneys in experimental animals (Gupta and Xie 2018). Further, with higher doses they may cross blood–brain barrier that may cause neurological damage (Oberdörster et al. 2004). Therefore, long-term experiments are needed to know further. The nanofertilizer particles may accumulate in soil and have the capability to disrupt soil ecosystems when recommended doses are not applied. Leaching of nanofertilizer particles may hamper aquatic organisms, which resulted in bioaccumulation and biomagnification within food chain (Nowack and Bucheli 2007). They may affect the non-target organisms including insects, fish, and birds (Handy et al. 2008). Therefore, more research is needed on ecological aspects before global acceptance of nanofertilizers.

## 1.10 Environmental Concern and Regulatory Considerations

An estimated 15–33% loss of nanofertilizer particles to the environment during plant application has been noticed. Moreover, with the application of higher concentration than recommended, it may induce toxic effect to plants such as cell proliferation, necrosis, apoptosis, DNA damage as well as oxidative stress. Therefore, more research work is needed for additivity, synergistic, potentiation, and antagonistic effects. That must be including into regulatory framework and policies so that it may protect human health with the application in agriculture. The other concerns like economic viability, scalability, social acceptability, knowledge gap, and global regulatory issues are very important to consider before its commercialization. They may also inhale to farm workers during spray which may not only reach to blood as well as other target site like brain, liver, and heart. Therefore, there must be planned program that can be implemented in case of soil, atmosphere, aquatic life, and food products exposed to the toxicity level of nanofertilizers. We have to assure about the sufficiency of current toxicity protocol and modify it, if needed. Detailed studies are needed about the impact of nanofertilizer particles within the human body after exposed through nanofood. The suggested risk management programs are: (1) Evaluating the hazard posed by the nanofertilizers based on toxicology and health effect data besides physical and chemical properties. (2) The degree of risk has been determined after assessing potential worker exposure data. (3) Education and training of handling of nanofertilizer to the worker. (4) Proper selection of protective equipment like clothing, gloves, respirators, etc. (5) Engineering control like exhaust ventilation at manufacturing unit should be regularly monitored. (6) Workers should be provided with the appropriate personal protective measures. Nanofertilizer may affect through respiration, ingestion, and skin. Dermal absorption of nanoparticles is merely investigated and urgently needs more systematic research. Educating workers or farmers on the safe handling and application of the fertilizer nanoparticles is very important before application in large scale that provides adequate safety to handle each product. Proper guidance should be given for waste disposal and cleaning of cloths as well as work areas. A guideline should be prepared for quality control of nanofertilizer products coming from different industries with the approval of the government. Figure 1.4 indicates issues to be considered before application of nanofertilizer globally.

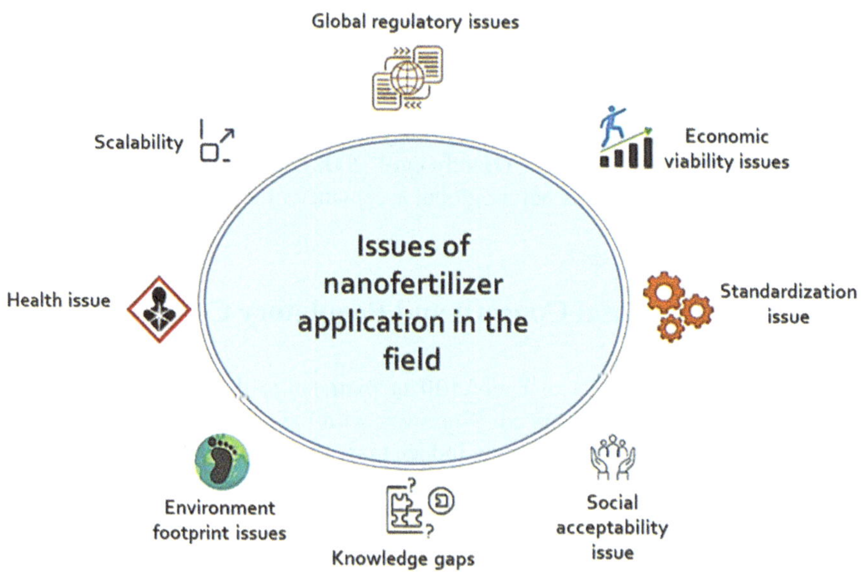


**Fig. 1.4** Matter in question of nanofertilizer application in the field

## 1.11 Future Prospects

Agricultural application of nanofertilizer provides an opportunity to attain the sustainability towards global food production and minimize hunger index. Nanosize fertilizer particles assured nutrient delivery to plants, increasing crop productivity, and minimizing the environmental impact of conventional fertilizers. It has been noted that various nanofertilizer materials like metal oxide nanoparticles, nanozeolites, nanochitosan, etc. have the potential to increase nutrient absorption and retention in soils. The controlled release nanofertilizers are already successfully developed which gradually supply nutrients minimizing the number of applications as well as nutrient loss. But more research should focus on ecofriendly and cost-effective synthesis method, and we need more green technology innovation methods for nanofertilizer production. More efforts are needed to produce smart nanofertilizers which can control release and targeted delivery of nutrient as and when needed by plants and that can help to minimize the nutrient losses and increase their use efficiency as well as may be ideal for sustainable agricultural practices. There is need to integrate nanosensor into nanofertilizers to monitor soil nutrient level, precise application and avoiding nutrient loss that way farmers may be benefited to use correct fertilizer in appropriate time and environmental pollution reduced considerably. Elaborate research also needed to know potential risk of applied nanofertilizers both to the environment and human health. There is an urgent need from the government side for clear regulatory framework and standardized guidelines to make certain of secure and effective use. Long-term studies should be initiated for potential toxicity of nanofertilizers on soil organisms, crops, and animals including humans. More research is needed on bionanoformulations, nanofertilizer use on seed treatment, nanofertilizer for precision farming, nanosensor, nanopesticides, and herbicides and their use for water management, food industry, and climate resilient. Immediate attention is needed to teach farmers about nanofertilizer use and disseminate the present knowledge to them regarding the benefits and safety as well as efficient use. A holistic evaluation is needed for life cycle assessment of various nanofertilizer particles and their effect on food chain. Their benefit should be validated in practical field settings to address stakeholders concerns. More research is needed to address the environmental fate, toxicity, and biodegradability of nanofertilizers. Moreover, in-depth risk assessment is needed to ensure safe and sustainable use.

## 1.12 Conclusion

Nanofertilizers turn up to be an excellent material to cope up with food security and reducing global hunger index due to their more nutrient use efficiency, native nutrient mobilization, controlled release of nutrients, targeted delivery, plant growth stimulation, and reducing environmental impacts as well as water and nutrient loss.

Foliage application is the most effective way to apply nanofertilizer to mobilize more nutrients from native sources by plant enzymes. In spite of potential benefit, more research is needed to correct nanomaterial composition, their safe application doses and release rate as well as to develop environment-friendly cost-effective scalable production methods. Long-term research is needed to find out the potential risk of nanofertilizer application on human, environmental and ecological systems. Therefore, strict guidelines are needed for their safe use to the farmers and quality production to the industries. There is a need for interdisciplinary collaboration among the researchers, policymakers, and agricultural experts for emerging, developing, and implementing nanofertilizers. The knowledge exchange among the policymakers and researchers as well as knowledge dissemination to the farmers is absolutely needed for global implications. There is an urgent need to educating the farmers about the advantages, application, safety measure as well as potential risks of nanofertilizers and ways to overcome them for more production and wide application. The refining of delivery mechanism should be fine tuned in future to minimize ecological footprint and increases in global security. Nanofertilizer technology is definitely unique and useful for modern agriculture and food security provided we address on potential risk, educate the farmers, and foster interdisciplinary collaboration to overcome the hurdles.

## References

Achari GA, Kowshik M (2018) Recent developments on nanotechnology in agriculture: plant mineral nutrition, health, and interactions with soil microflora. J Agric Food Chem 66:8647–8661

Ahmed S, Fahmy A (2017) Environmental impact for applications of neem cake coated urea and nano iron foliar on rationalization of chemical nitrogen fertilizers and wheat yield. J Soil Sci Agric Eng 8:613–620

Aslani F, Bagheri S, MuhdJulkapli N, Juraimi AS, Hashemi FS, Baghdadi A (2014) Effects of engineered nanomaterials on plants growth: an overview. Sci World J 641759:1–28

Aththanayaka S, Thiripuranathar G, Ekanayake S (2022) Emerging advances in biomimetic synthesis of nanocomposites and potential applications. Mater Today Sustain 20:100206

Babu S, Singh R, Yadav D, Rathore SS, Raj R, Avasthe R, Yadav SK, Das A, Yadav V, Yadav B et al (2022) Nanofertilizer for agricultural and environmental sustainability. Chemosphere 292:133451

Chhipa H, Joshi P (2016) Nanofertilisers, nanopesticides and nanosensors in agriculture. In: Nanoscience in food and agriculture 1. Springer, Berlin/Heidelberg, pp 247–282

Davarpanah S, Tehranifar A, Davarynejad G, Abadía J, Khorasani R (2016) Effects of foliar applications of zinc and boron nano-fertilizers on pomegranate (*Punica granatum* cv. Ardestani) fruit yield and quality. Sci Hortic 210:57–64

DeRosa MC, Monreal C, Schnitzer M, Walsh Y, Sultan Y (2010) Nanotechnology in fertilizers. Nat Nanotechnol 5:91

Fatima F, Hashim A, Anees S (2021) Efficacy of nanoparticles as nanofertilizer production: a review. Environ Sci Pollut Res 28:1292–1303

Gupta R, Xie H (2018) Nanoparticles in daily life: applications, toxicity and regulations. J Environ Pathol Toxicol Oncol 37:209–230

Handy RD, Owen R, Valsami-Jones E (2008) The ecotoxicology of nanoparticles and nanomaterials: current status, knowledge gaps, challenges, and future needs. Ecotoxicology 17:315–325

Jakhar AM, Aziz I, Kaleri AR, Hasnain M, Haider G, Ma J, Abideen Z (2022) Nano-fertilizers: a sustainable technology for improving crop nutrition and food security. NanoImpact 27:100411
Kalia A, Kaur H (2019) Nano-biofertilizers: harnessing dual benefits of nano-nutrient and bio-fertilizers for enhanced nutrient use efficiency and sustainable productivity. Nanosci Sustain Agric:51–73
Kaushal M, Wani SP (2017) Nanosensors: frontiers in precision agriculture. In: Nanotechnology. Springer, Berlin/Heidelberg, pp 279–291
Khodakovskaya M, Dervishi E, Mahmood M, Xu Y, Li Z, Watanabe F, Biris AS (2009) Carbon nanotubes are able to penetrate plant seed coat and dramatically affect seed germination and plant growth. ACS Nano 3:3221–3227
KhordadiVaramin J, Fanoodi F, Sinaki JM, Rezvan S, Damavandi A (2018) Physiological response of sesame (*Sesamum indicum* L.) to application of chitosan and magnesium-nano fertilizers under irrigation cut-off in a sustainable agriculture system. Iran J Plant Physiol 9:2629–2639
Kong X-P, Zhang B-H, Wang J (2021) Multiple roles of mesoporous silica in safe pesticide application by nanotechnology: a review. J Agric Food Chem 69:6735–6754
Kottegoda N, Munaweera I, Madusanka N, Karunaratne V (2011) A green slow-release fertilizer composition based on urea-modified hydroxyapatite nanoparticles encapsulated wood. Curr Sci 101:73–78
Krishnamoorthy R, Gassem MA, Athinarayanan J, Periyasamy VS, Prasad S, Alshatwi AA (2021) Antifungal activity of nanoemulsion from Cleome viscosa essential oil against food-borne pathogenic *Candida albicans*. Saudi J Biol Sci 28:286–293
Krishnani KK, Boddu VM, Chadha NK, Chakraborty P, Kumar J, Krishna G, Pathak H (2022) Metallic and non-metallic nanoparticles from plant, animal, and fisheries wastes: potential and valorization for application in agriculture. Environ Sci Pollut Res 29:81130–81165
Kubavat D, Trivedi K, Vaghela P, Prasad K, Vijay Anand GK, Trivedi H, Patidar R, Chaudhari J, Andhariya B, Ghosh A (2020) Characterization of a chitosan-based sustained release nanofertilizer formulation used as a soil conditioner while simultaneously improving biomass production of *Zea mays* L. Land Degrad Dev 31:2734–2746
Lehmann J, Kleber M (2015) The contentious nature of soil organic matter. Nature 528:60–68
Liao YY, Huang Y, Carvalho R, Choudhary M, Da Silva S, Colee J, Huerta A, Vallad GE, Freeman JH, Jones JB et al (2021) Magnesium oxide nanomaterial, an alternative for commercial copper bactericides: field-scale tomato bacterial spot disease management and total and bioavailable metal accumulation in soil. Environ Sci Technol 55:13561–13570
Lin N, Huang J, Chang PR, Anderson DP, Yu J (2011) Preparation, modification, and application of starch nanocrystals in nanomaterials: a review. J Nanomater 2011:1–13
Liu R, Lal R (2015) Potentials of engineered nanoparticles as fertilizers for increasing agronomic production. Sci Total Environ 514:131–139
Mejias J, Salazar F, Pérez Amaro L, Hube S, Rodriguez M, Alfaro M (2021) Nanofertilizers: a cutting-edge approach to increase nitrogen use efficiency in grasslands. Front Environ Sci Eng 9:52
Mujtaba M, Khawar KM, Camara MC, Carvalho LB, Fraceto LF, Morsi RE, Elsabee MZ, Kaya M, Labidi J, Ullah H et al (2020) Chitosan-based delivery systems for plants: a brief overview of recent advances and future directions. Int J Biol Macromel 154:683–697
Nowack B, Bucheli TD (2007) Occurrence, behavior and effects of nanoparticles in the environment. Environ Pollut 150:5–22
Oberdörster G, Sharp Z, Atudorei V, Elder A, Gelein R, Kreyling W, Cox C (2004) Translocation of inhaled ultrafine particles to the brain. Inhal Toxicol 16:437–445
Raliya R, Tarafdar JC, Mahawar H, Kumar R, Gupta P, Mathur T, Kaul RK (2014) ZnO nanoparticles induced exopolysaccharide production by *B. subtilis* strain JCT 1 for arid soil applications. Int J Biol Macromol 65:362–368
Raliya R, Saharan V, Dimkpa C, Biswas P (2018) Nanofertilizer for precision and sustainable agriculture: current state and future perspectives. J Agric Food Chem 66:6487–6503

Ramirez-Rodriguez GB, Dal Sasso G, Carmona FJ, Miguel-Rojas C, Perez-de-Luque A, Masciocchi N, Guagliardi A, Delgado-Lopez JM (2020) Engineering biomimetic calcium phosphate nanoparticles: a green synthesis of slow-release multinutrient (NPK) nanofertilizers. ACS Appl Bio Mater 3:1344–1353

Satti SH, Raja NI, Ikram M, Oraby HF, Mashwani Z-U, Mohamed AH, Singh A, Omar AA (2022) Plant-based titanium dioxide nanoparticles trigger biochemical and proteome modifications in *Triticum aestivum* L. under biotic stress of *Puccinia striiformis*. Molecules 27:4274

Saxena M, Maity S, Sarkar S (2014) Carbon nanoparticles in 'biochar' boost wheat (Triticum aestivum) plant growth. RSC Adv 4:39948–39954

Sempeho SI, Kim HT, Mubofu E, Hilonga A (2014) Meticulous overview on the controlled release fertilizers. Adv Chem 2014:1–16

Sharma U, Barupal M, Shekhawat NS, Kataria V (2018) Aeroponics for propagation of horticultural plants: an approach for vertical farming. Hortic Int J 2:443–444

Sharma V, Javed B, Byrne H, Curtin J, Tian F (2022) Zeolites as carriers of nano-fertilizers: from structures and principles to prospects and challenges. ACS Appl Nanomater 3:163–186

Sharma B, Tiwari S, Kumawat KC, Cardinale M (2023) Nano-biofertilizers as bio-emerging strategies for sustainable agriculture development: potentiality and their limitations. Sci Total Environ 860:160476

Sheoran P, Goel S, Boora R, Kumari S, Yashveer S, Grewel S (2021a) Biogenic synthesis of potassium nanoparticles and their evaluation as a growth promoter in wheat. Plant Gene 27:100310

Sheoran P, Grewal S, Kumari S, Goel S (2021b) Enhancement of growth and yield, leaching reduction in Triticum aestivum using biogenic synthesized zinc oxide nanofertilizer. Biocatal Agric Biotechnol 32:101938

Shilpa RS, Kant C, Prashar N (2022) Role of nanofertilizers in horticulture: a review. J Pharm Innov 11:831–836

Singh P, Singh R, Verma P, Bhadouria R, Kumar A, Kaushik M (2021) Plant-microbes-engineered nano-particles (PM-ENPs) nexus in agro-ecosystems: understanding the interaction of plant, microbes and engineered nano-particles (ENPS). Springer, Berlin/Heidelberg

Tang Q, Xu Z, Hong A, Zhang X, Kah M, Li L, Wang Y (2021) Response of soil enzyme activity and bacterial community to copper hydroxide nanofertilizer and its ionic analogue under single versus repeated applications. Sci Total Environ 796:148974

Tarafdar JC (2021) Nanofertilizers: challenges and prospects. Scientific Publisher (India), p 363

Tarafdar JC (2022) Nanotechnology application in agriculture: current developments of nanofertilizers. In: Singh P, Anam, Srivastava TK, Verma RR (eds) Chapter 2—Nanoparticles application in agriculture. Scientific Publishers, pp 19–29

Tarafdar JC, Xiang Y, Wang WN, Dong Q, Biswas P (2012) Standardization of size, shape and concentration of nanoparticle for plant application. Appl Biol Res 14:138–144

Tarafder C, Daizy M, Alam MM, Ali MR, Islam MJ, Islam R, Ahommed MS, Aly AS, M., & Khan, M.Z.H. (2020) Formulation of a hybrid nanofertilizer for slow and sustainable release of micronutrients. ACS Omega 5:23960–23966

Thirugnanasambandan T (2021) Advances of engineered nanofertilizers for modern agriculture. In: Plant-microbes-engineered nano-particles (PM-ENPs) nexus in agro-ecosystems. Springer, Berlin/Heidelberg, pp 131–152

Thiruvengadam M, Rajakumar G, Chung IM (2018) Nanotechnology: current uses and future applications in the food industry. Biotechnology 8:74

Umarani R, Mala R (2013) Influence of calcium phosphate nano gel fertilizer composite on enzymes, biomolecules and yield of Abelmoschus esculentus. Int J Agric Environ Biotechnol 6:771

Vejan P, Khadiran T, Abdullah R, Ahmad N (2021) Controlled release fertilizer: a review on developments, applications and potential in agriculture. J Control Release 339:321–334

Wilson MA, Tran NH, Milev AS, Kannangara GK, Volk H, Lu GM (2008) Nanomaterials in soils. Geoderma 146:291–302

Zhang X, Cao H, Wang H, Zhao J, Gao K, Qiao J, Li J, Ge S (2022) The effects of graphene-family nanomaterials on plant growth: a review. Nano 12:936

# Chapter 2
# Nanotechnology in Agriculture: Harnessing Nanofertilizers for Sustainable Global Food Security

**Ashok K. Patra and Vassanda Coumar M**

**Abstract** The extensive misuse of chemical fertilizers in contemporary agricultural practices has led to soil degradation and environmental pollution, necessitating the adoption of precision fertilizer management for sustainable agriculture and nutritional security. Nanofertilizers emerge as innovative solutions to revolutionize nutrient management, enhance crop yield, and promote environmental sustainability. By utilizing nano-sized materials for controlled and targeted nutrient release, nanofertilizers improve nutrient use efficiency, reduce fertilizer requirements, and minimize environmental pollution compared to traditional fertilizers. Traditional fertilizers often suffer from inefficiencies such as nutrient leaching and limited uptake by plants, leading to over-fertilization and suboptimal crop growth. In contrast, nanofertilizers offer controlled nutrient release, targeted delivery, and increased bioavailability, enhancing nutrient use efficiency and improving soil fertility. The unique properties of nanomaterials enable interactions with soil particles, influencing nutrient adsorption and surface interactions. Nanofertilizers encompass nanoparticles, nanocomposites, and nanocapsules, each offering unique advantages in nutrient delivery and controlled release. Experimental studies demonstrate the efficacy of nanofertilizers in various crops, enhancing phosphorus uptake, reducing nitrogen losses, and improving crop productivity. Furthermore, nanofertilizers contribute to sustainable water management by enhancing water retention and reducing water requirements through hydrogel incorporation and seed coatings. Integration of nanofertilizers with precision agriculture optimizes nutrient delivery, enhances soil structure, and improves crop quality, ensuring sustainable and resilient farming systems. Despite their promising benefits, challenges such as environmental and health concerns, high production costs, and regulatory standards need to be addressed for the responsible and sustainable deployment of nanofertilizers. Future research directions include improving fabrication processes, understanding uptake mechanisms, and addressing toxicity concerns to ensure the safe and effective implementation of nanofertilizers in agriculture. Overall, nanofertilizers offer a

A. K. Patra (✉) · V. Coumar M
ICAR-Indian Institute of Soil Science, Bhopal, Madhya Pradesh, India



A. K. Patra (ed.), *Nanofertilizers in Agriculture*, Sustainability Sciences in Asia and Africa, https://doi.org/10.1007/978-981-97-8112-6_1

transformative potential to address global food security challenges while promoting environmental sustainability and efficient resource utilization in agriculture.

**Keywords** Nanofertilizers · Nanotechnology · Precision fertilizer management · Sustainable agriculture · Nutrient management · Environmental sustainability · Nutrient use efficiency

## 2.1 Introduction

Global food security faces various challenges arising from a complex interaction of social, economic, environmental, and political factors. Based on the present world's population and its ever-increasing growth, it is projected to reach around 9.7 billion people by 2050 (FAO 2018). Accordingly, it has been anticipated that current crop production needs to be increased by up to 70% to satisfy future food demands (Hunter et al. 2017). However, the limited availability of essential resources such as arable land, water, and energy poses a formidable constraint on expanding agricultural production. The pressures intensify with urbanization and industrialization, exacerbating competition for these resources and compelling agricultural systems to augment their output. Additionally, climate change-induced shifts in weather patterns, manifesting as extreme events, droughts, floods, and temperature variations, disrupt traditional growing seasons, leading to crop failures and diminishing agricultural productivity. To effectively address these challenges and ensure global food security, comprehensive strategies are imperative. These strategies must acknowledge the interconnected nature of the issues at hand and involve the implementation of sustainable agricultural practices, technological innovations, and international cooperation.

As the agricultural sector grapples with the imperative of sustaining a burgeoning global population, nanotechnology emerges as a promising frontier, offering innovative solutions for sustainable and precision agriculture. Recognizing its potential applications in revolutionizing traditional farming practices, researchers and scientists are increasingly turning to nanotechnology as a tool to enhance agricultural productivity and tackle food security challenges. While nanotechnology has achieved significant breakthroughs in medicine and pharmacology, its applications in agriculture have garnered comparatively less attention. The fundamental building blocks of nanotechnology, nanoparticles, exhibit unique properties distinct from bulk materials, such as increased surface area and altered reactivity. These properties make nanotechnology particularly advantageous in various applications, spanning medicine, energy, and notably agriculture.

The application of nanotechnology in agriculture, often referred to as nanobiotechnology or nanobiology, aims to improve different facets of farming practices, including crop yield, nutrient management, pest control, and soil health. Nanomaterials' unique properties enable targeted and controlled delivery of substances, paving the way for more efficient and sustainable agricultural processes

(Liu and Lal 2015; DeRosa et al. 2010). Nanomaterials find applications in agriculture as growth regulators, nanopesticides, nanofertilizers, nanoantimicrobial agents, and carriers for target genes in plants (Raimondi et al. 2021; Monreal et al. 2016; Raliya et al. 2018).

Furthermore, nanotechnology holds promise as a strategy to mitigate abiotic stress in plants, contributing to sustainable agriculture. Nanosensors and nanodevices play a pivotal role in precision agriculture, offering real-time data on soil conditions, crop health, and environmental factors (Fig. 2.1). This data-driven approach empowers farmers to optimize resource use through informed decision-making. To fully realize these benefits and address potential challenges responsibly, continuous research, education, and collaboration remain essential. Only through concerted efforts we can unlock the full potential of nanotechnology and steer global agriculture toward a more secure and sustainable future. This article outlines conventional, modern, and nanomaterial-based agricultural practices, emphasizing the advantages of nano-based approaches for targeted agrochemical delivery, particularly the fertilizers, the biosensor integration, and precision farming.

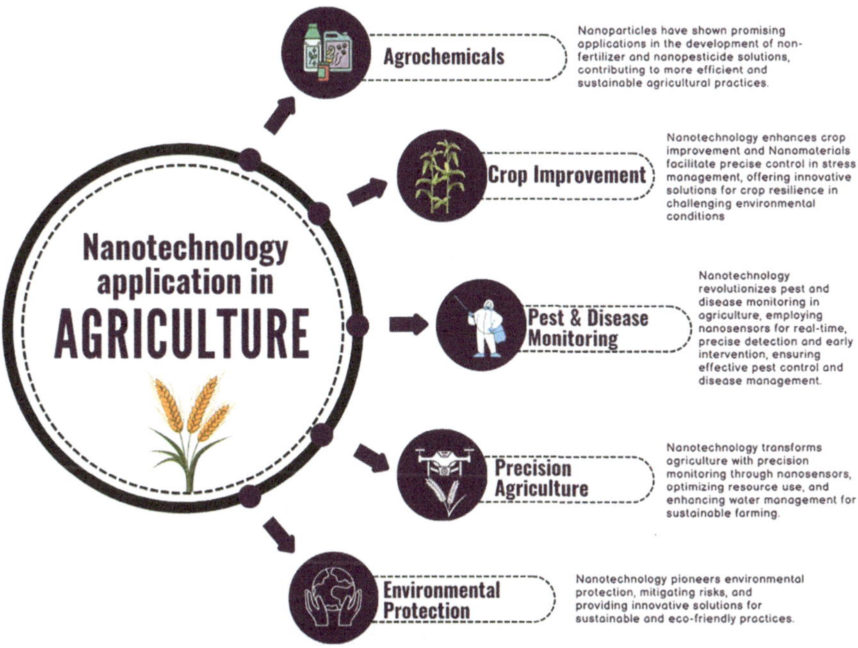


**Fig. 2.1** Nanotechnological application in agriculture for sustainable ecosystem production

## 2.2 Enhanced Nutrient Use Efficiency

The extensive misuse of chemical fertilizers in contemporary agricultural practices has wreaked havoc on once-fertile soils, depleting vital nutrients and disrupting the delicate balance of the surrounding ecosystem. Precision fertilizer management stands as a pivotal requirement for sustainable agricultural development and nutritional security, given the projected population growth and the need for a 70% increase in crop production by 2050 (FAO 2018; Hunter et al. 2017). In response to these challenges, nanofertilizers emerge as innovative solutions with the potential to revolutionize nutrient management, enhance crop yield, and promote environmental sustainability. The strategic implementation of nanofertilizers, aligned with sustainable and precision farming practices, holds immense promise for securing the long-term sustainability of global agricultural systems. Nano-sized materials, designed for controlled and targeted nutrient release, ensure that essential nutrients are efficiently delivered to plants, thus improving nutrient use efficiency. This not only reduces the overall amount of fertilizer required but also minimizes nutrient losses, contributing to environmental pollution.

Traditional fertilizers often suffer from inefficiencies related to nutrient leaching, volatilization, and limited uptake by plants. Traditional fertilizers are often applied uniformly across fields, regardless of variations in soil nutrient levels. Such imprecise practices can result in over-fertilization in certain areas and under-fertilization in others, leading to suboptimal crop growth. Additionally, traditional fertilizers often provide an immediate nutrient burst, which may not align with the plant's growth requirements, leading to wastage and inefficient resource utilization. Nitrogen-based fertilizers, in particular, have caused significant nitrogen losses into the environment, contributing to water body eutrophication, soil acidification, and biodiversity decline (Banger et al. 2017). The loss of nutrients from the soil has significantly impacted soil fertility, with traditional fertilizers exhibiting low nutrient efficiency (30–50% N, 15–20% P, 60–70% K, 8–10% S, and 1–2% micronutrients). In contrast, nanofertilizers, owing to their nanoscale dimensions, offer controlled nutrient release, targeted delivery, and increased bioavailability, ultimately enhancing nutrient use efficiency (Yadav et al. 2023).

The unique properties of nanomaterials enable them to interact with soil particles, influencing adsorption, surface interactions, and undergoing processes like agglomeration, sedimentation, or dissolution. Environmental factors, including temperature, pH, and moisture, play a role in nutrient liberation from nanofertilizers, while soil microorganisms contribute to their transformation. This intricate interplay of factors underscores the need for assessing the long-term impacts of nanofertilizers on agricultural production systems and environmental protection (Ghormade et al. 2011). The increased nutrient use efficiency facilitated by nanofertilizers represents a significant advancement in nutrient management strategies, offering a more precise, environmentally friendly, and economically viable approach to fertilization (Liu and Lal 2015; DeRosa et al. 2010).

Nanotechnology involves the manipulation and control of materials at the nanoscale, typically ranging from 1 nm to 100 nm (Lee and Moon 2020). At this level, the unique physical, chemical, and biological properties of materials emerge, allowing for precise engineering and manipulation. The prefix "nano" denotes one billionth of a meter, highlighting the minuscule scale at which these technologies operate. Reducing the size to the nanoscale enhances the availability of nutrients to plant pores, promoting efficient nutrient uptake (Suppan 2013). Nanoparticles (NPs) can infiltrate plant cells and transport nutrients by binding to carrier proteins, utilizing aquaporins, ion channels, endocytosis, or attaching to organic chemicals present in plant tissues (Rico et al. 2011). Nanoparticles that occur naturally can be sourced from a variety of origins, encompassing volcanic ash, the ocean, and biological substances like viruses and dust. Nanofertilizers are diverse and can be classified into different types based on their composition and structure. The three main categories are nanoparticles, nanocomposites (nanoscale additives), and nanocapsules (nanoscale coatings) (Mikkelsen 2018). Each type offers unique advantages in terms of nutrient delivery, controlled release, and targeted action. Nanoparticles are synthesized using various materials including metals (such as zinc, iron, or copper), metal oxides, and carbon-based materials. Carbon-based nanofertilizers generally utilizes carbon nanotubes, graphene, or other carbon-based materials as carriers for nutrients. These nanoparticles exhibit unique properties, including a high surface area-to-volume ratio and enhanced reactivity, which contribute to improved nutrient solubility and bioavailability (Gaur et al. 2021). Similarly, nanocomposites are hybrid materials composed of two or more different types of nanoscale components, often combining nanoparticles with a matrix material. Nanocapsules are nanoscale containers or carriers that encapsulate active substances, protecting them from environmental factors until release is triggered. The shell of nanocapsules is typically composed of biodegradable polymers or lipid-based materials, while the core contains the active ingredients, such as nutrients or growth-promoting substances. Nanocapsules provide a protective barrier around the encapsulated substances and facilitates controlled release of nutrient to plant (Borges et al. 2019; Tarafder et al. 2020; Golbashy et al. 2016). This controlled release mechanism enhances nutrient efficiency and reduces environmental losses. Thus, understanding the different types of nanofertilizers allows researchers and farmers to choose or design formulations that best suit their agricultural needs. The versatility of nanofertilizers offers a promising avenue for addressing challenges in traditional fertilization practices and promoting sustainable agriculture.

Numerous experimental studies have explored the efficacy of nanofertilizers for soil or foliar applications in different plant species. Porous nanomaterials, such as zeolites, clay, or chitosan, have demonstrated success in reducing nitrogen losses by regulating demand-based release and enhancing plant uptake (Millan et al. 2008; Abdel-Aziz et al. 2016). Studies using Isotope ($^{33}P$) dilution techniques have revealed enhanced phosphorus uptake and utilization in wheat crops treated with Hydroxyapatite nanoparticles (n-HAP) compared to bulk-HAP (Montalvo et al. 2015). Coating traditional fertilizers like monoammonium phosphate and urea with nano-ZnO and nano-sulfur has resulted in gradual nutrient release, improving zinc

(Zn) and sulfur (S) uptake efficiency (Wilson et al. 2008; Milani et al. 2012). Likewise, the use of hydroxyapatite nanoparticles to coat and encapsulate urea into softwood cavities doubled the nitrogen release compared to conventional urea fertilizer, as reported by Kottegoda et al. (2011). Ammonium-zeolites and potassium-graphene oxide NFs, improved the efficiency of phosphorus (P) and potassium (K) utilization by preventing leaching and ensuring the precise release of nutrients according to the plant nutrient demand. Nano-calcite ($CaCO_3$) incorporation with nanoparticles like $Fe_2O_3$, $SiO_2$, and MgO has elevated the absorption of various nutrients, showcasing the versatility of nanofertilizers in comparison to soluble fertilizers (Sabir et al. 2014a, b). These controlled-release mechanisms of nanofertilizers prove to be a remarkable alternative, releasing nutrients gradually, allowing efficient utilization by plants throughout the entire crop growth period, minimizing nutrient loss, and improving use efficiencies (Singh et al. 2019). Thus, nanofertilizers emerge as a revolutionary alternative to conventional fertilization practices, providing controlled nutrient release, targeted delivery, and enhanced bioavailability. The ongoing research and application of nanofertilizers in diverse agricultural contexts are imperative for fully realizing their potential and advancing sustainable agriculture on a global scale.

## 2.3 Sustainable Water Management Through Nanotechnology

Efficient water management is crucial in agriculture to promote crop growth, maximize yield, and optimize the utilization of nutrients and other resources. Traditionally, farmers have relied on methods such as visual inspection and manual soil sampling to assess soil moisture levels. However, these methods are time-consuming, prone to human error, and lack real-time data. With increasing weather unpredictability and the need for sustainable farming, accurate and prompt soil moisture monitoring has become essential. Researchers are actively developing strategies to enhance water use efficiency in agriculture, such as advanced drip irrigation systems. These systems focus on critical factors including water storage, in situ water holding capacity, water distribution near roots, plant water absorption efficiency, encapsulated water release on demand, and integration with field intelligence through technological advancements. To address these challenges, nanofertilizers offer a promising solution for improving irrigation efficiency by enhancing water retention and reducing water requirements. This is particularly important in water-scarce regions, promoting sustainable water management practices and boosting overall crop productivity.

Nanofertilizers significantly contribute to water conservation by improving water efficiency across various applications. They enhance the soil's water absorption and retention capabilities through the use of nanomaterials like hydrogels—tiny particles with high water-absorbing capacities. Incorporating hydrogels into the soil increases its moisture content, allowing more efficient water access for plants and reducing the need for irrigation. Additionally, applying nanomaterial-based coatings to seeds improves water absorption, leading to better germination rates and overall plant growth. For example, nanofertilizers like urea coated with nanoparticles of Cu, Zn, and FeO are designed to reduce water loss by making the fertilizer surface hydrophilic. This modification enhances water-holding capacity and minimizes water evaporation (Jakhar et al. 2022). Specific nanofertilizers are also engineered to improve water use efficiency by enhancing the plant's ability to absorb and utilize water. Nanoparticles of clay and carbon, for instance, can bind soil particles, forming larger aggregates that improve soil structure. This enhancement increases water-holding capacity and reduces soil erosion (Raliya et al. 2018).

Furthermore, the integration of nanofertilizers can lead to significant gains in water efficiency, thereby reducing the energy needs associated with water transport and use. The US Environmental Protection Agency (2016) reported that up to 50% of water used for irrigation is wasted due to inefficient methods and systems. Innovations such as nanoscale coatings with self-healing properties could repair leaky pipes, potentially reducing friction loss by 50% and minimizing energy requirements for water transportation. Nanoemulsions also play a crucial role in controlling soil water loss (Mali et al. 2020). In a study by Al-Shaabani et al. (2023), the impact of nanotechnology-enhanced perfusion irrigation, along with surface and subsurface drip irrigation, on irrigation efficiency was investigated. The nanotechnology perfusion irrigation method resulted in the lowest seasonal water addition, achieving notable water savings of 29.97% and 13.19% compared to surface drip irrigation, and 15.23% and 8% compared to subsurface drip irrigation in plowed and non-plowed soils, respectively. Moreover, plants equipped with various nanosensors help in scheduling irrigation and crop inputs based on crop demand. Mobile sensors predict various stresses, enabling timely and precise management. This integration of nanofertilizers into water management practices has the potential to revolutionize agricultural methods, ensuring efficient resource utilization and contributing to sustainable and resilient farming systems.

## 2.4 Nanofertilizers in Enhancing Crop Yield and Quality

Nanofertilizers hold significant promise for increasing crop production and improving crop quality while preserving the environment (Yata et al. 2017). The physical and chemical characteristics of nanofertilizers, such as size, type, composition, and application method, play a crucial role in the growth and development of crop plants. Recent studies demonstrate that nanofertilizers, including silver nanoparticles (NPs), ZnO NPs, silica NPs, and $TiO_2$ NPs, stimulate seed germination and plant growth in various cereal and vegetable crops, such as wheat, rice, maize, sorghum, tomato, onion, and radish. These nanofertilizers contribute to increased crop quality and productivity (Parveen et al. 2016; Maroufpoor et al. 2019; Mirzajani et al. 2013).

Numerous investigations reveal that nanofertilizers enhance water absorption and penetration in the seed coat, improving seed germination and seedling growth by stimulating the enzyme system (Abou-Zeid et al. 2020). Kah et al. (2018) reported that nanofertilizers can enhance crop production by up to 30% compared to traditional fertilizers, influencing the physiology and biochemistry of plants. The reduction in particle size facilitates effective adsorption of nutrient ions on soil particles, resulting in a slower release of nutrient ions aligned with crop nutrient demand, ultimately improving Nutrient Use Efficiency (NUE) and crop productivity. For instance, the application of ZnO NPs increased barley productivity by 60% compared to a control, surpassing the 31% increase achieved with $ZnSO_4$ fertilizer in Zn-deficient soil (Kale and Gawade 2016; Kale and Kane 2017). Silver nanoparticles (SNPs) were found to stimulate wheat growth and crop yield, leading to an 11 to 29% increase in rice, wheat, soybean, and maize yield when carbon dots were applied with traditional fertilizers (Iqbal and Umar 2019). Serval studies highlight the superiority of nanofertilizers (NFs) over conventional fertilizers in enhancing the growth, yield, and fruit quality of various crops (Sabir et al. 2014a, b; Song et al. 2016). Liu et al. (2009) reported a 20% increase in grain yields in vegetable crops with the application of carbon nanoparticles along with conventional fertilizers. Nano copper oxide demonstrated a 16–19% increase in lettuce crop biomass and elevated copper levels in the shoot, while conventional CuO did not exhibit similar enhancements (Wang et al. 2019). Foliar spray of Se-NPs in pomegranate significantly increased fruit production in terms of number, diameter, and peel compared to the control (Zahedi et al. 2019). Application of nanofertilizers–NPKMg fertilizer spray at 0.1% to mango shoots and leaves increased fruit production by 56% compared to conventional macronutrient (Saied 2018).

Nanofertilizers not only enhance crop yield but also have the potential to improve the nutritional content of harvested crops, addressing both the quantity and quality of the food produced. Several research attempts were made to improve nutritional (dietary) value and to extend the shelf-life of fruits and food grain using nanotechnology. Application of nanoparticle (nano-TiO) significantly increased the mineral uptake (P, Se, Mn, Fe, Mg, Ca, and Cu) and enhanced the crop quality of basil (Ocimum basilicum) plants (Tan et al. 2017). Fruit quality in terms of its firmness and vitamin content has been improved through foliar application copper nanoparticles in tomatoes, thus enhancing shelf life, taste and nutritional value of the fruit (López-Vargas et al. 2018). Kole et al. (2013) observed that nanofertilizers in bitter melon cultivation resulted in noticeable enhancements in fruit size, quantity, and quality, evident through increased levels of bioactive compounds such as cucurbitacin-B, lycopene, charantin, and inulin content. The foliar application of nano Zn and B in pomegranate resulted in a notable increase in fruit productivity ranging from 21% to 46%. Additionally, it contributed to improved fruit quality, including elevated levels of total soluble solids (4–8%), an increased maturity index (21–46%), higher juice pH (0.3–0.6 units), Zn content (61%), B content (20%) and a reduction in titratable acidity (10% to 29%) (Davarpanah et al. 2016). Kamiab and Zamanibahramabadi (2016) also reported enhancement in nutrient element of Zn, Fe, Cu, and Mn in the NP treated almond trees as compared to conventional fertilizers. The application of nanofertilizers not only enhances the content of specific elements but also increases the content of other essential elements in crops. For instance, exposure to $TiO_2$ NPs in cucumber seedlings produced approximately 34% more P content and 35% more K content than those in the control (Servin et al. 2012). Dekkers et al. (2011) reported that $SiO_2$ NPs were used to improve fragrances in the food items. Numerous nanoparticles, including FeO, ZnO, CuO, and $TiO_2$, have enhanced the quality of crops such as wheat, mungbean, soybean, mustard, and vegetables, leading to both improved produce quality and increased crop yield (Dubey and Mailapalli 2016; Shojaei et al. 2019). Nanofertilizers with their nano-sized particles, provide a larger surface area for nutrient interaction with plant roots, ensure controlled release mechanisms, enable targeted nutrient delivery, and improved nutrient use efficiency. These properties contribute to better overall plant health, translating into enhanced crop yield and quality.

## 2.5 Integration of Nanofertilizers with Precision Agriculture

Unlike traditional input-intensive agricultural systems, precision farming strives for maximum crop yield with minimal use of fertilizers and other inputs. Nanofertilizers play a crucial role in transforming traditional agriculture into smart and precision agriculture, promoting energy efficiency and eco-friendliness for sustainable agricultural practices. By integrating global satellite positioning systems, computers, and remote sensing devices, precision agriculture optimizes agricultural inputs through detailed analysis of environmental conditions, assessing soil conditions, plant development, water usage, and other relevant parameters (Sekhon 2014). Nanosensors play a pivotal role in real-time monitoring of plant growth, forecasting soil environmental changes, and timely nutrient management. This capability enables farmers to respond effectively to changing environmental conditions and effectively managing soil nutrient status (Yata et al. 2017).

Nanofertilizers enhance precision farming by providing more efficient nutrient delivery, which improves crop yield and quality. Their nano-sized particles ensure a larger surface area for interaction with plant roots, enabling targeted nutrient delivery and controlled release mechanisms. This leads to better nutrient use efficiency, translating into enhanced plant health and productivity. For instance, nanofertilizers like urea coated with nanoparticles of Cu, Zn, and FeO improve the plant's ability to absorb and utilize nutrients, resulting in higher crop yields (Jakhar et al. 2022). Nanofertilizers also improve soil structure and nutrient availability. Nanoparticles of clay and carbon bind soil particles, forming larger aggregates that enhance soil structure, increase water-holding capacity, and reduce soil erosion. These improvements in soil quality contribute to better crop growth and higher yields (Duhan et al. 2017; Raliya et al. 2018). Moreover, the integration of nanofertilizers with precision agriculture technologies enables real-time monitoring and precise management of nutrient levels in the soil. This capability allows farmers to respond effectively to changing environmental conditions and optimize fertilizer application, reducing waste and enhancing crop productivity. For example, ZnO nanoparticles have been shown to increase barley productivity by 60% in Zn-deficient soils, significantly outperforming traditional ZnSO4 fertilizers (Kale and Gawade 2016; Kale and Kane 2017). Overall, nanofertilizers enhance agricultural output by providing precise and efficient nutrient delivery, improving soil structure, and enhancing the nutritional quality of crops. Their integration with precision agriculture practices ensures sustainable and resilient farming systems, contributing to increased crop yield and quality.

## 2.6 Nanofertilizers for Reduction of Environmental Footprint

To address the escalating global food demand, the need for a sustainable, productive system with reliable technologies ensuring environmental protection is evident (Jyothi and Hebsur 2017). The indiscriminate use of conventional fertilizers in highly intensive agriculture poses a significant threat, as a substantial portion of the chemicals remains in the soil, adversely affecting its quality. Additionally, these chemicals can infiltrate other environmental compartments such as air and water through groundwater leaching, surface runoff, and drift, causing severe environmental contamination that can disrupt the normal growth of flora and fauna.

Globally, efforts are being made through nanotechnology to reduce fertilizer consumption in agriculture without compromising crop productivity and environmental sustainability (Das et al. 2016). On a global scale, agriculture is a major contributor to atmospheric nitrogen, with over 50% of the nitrogen applied to agricultural soils potentially lost into the environment, contributing to global warming as a significant anthropogenic greenhouse gas to (Aneja et al. 2009; Davidson 2009). By minimizing nutrient losses to the environment, nanofertilizers helps to reduce the environmental footprint associated with fertilizer application, aligning with sustainable agriculture practices and mitigating the negative impacts of nutrient runoff on ecosystems. Certain nanofertilizers, designed to reduce nitrogen volatilization, can decrease greenhouse gas emissions, particularly nitrous oxide emissions associated with nitrogen fertilizers. The controlled release mechanisms of nanofertilizers, coupled with improved nutrient uptake efficiency, help minimize nutrient leaching into the soil, thus reducing groundwater contamination and eutrophication.

The development of nanofertilizers represents a significant step forward in environmental conservation, as they can be applied in smaller amounts compared to conventional fertilizers (Adisa et al. 2019). This reduction in application quantity diminishes the potential for leaching, runoff, and gas emissions into the atmosphere (Manjunatha et al. 2016). Nanofertilizers can be customized to specific soil and climatic conditions, ensuring optimal nutrient release. The precision and efficiency of nanofertilizers enable more accurate nutrient application, minimizing the likelihood of overapplication, a common cause of nutrient runoff. Reducing overapplication contributes to lower environmental impact. Reports indicate a reduction in nitrogen losses using N-charged porous nanomaterials like zeolites (Manikandan and Subramanian 2017), clays (Sarkar et al. 2014), or biodegradable polymers such as chitosan (Sharif et al. 2018). Nanofertilizers, such as $SiO_2$-N, remain fixed in the soil for an extended period, enhancing fertilizer use efficiency and preventing soil and water contamination, ensuring biosafety (Abbas et al. 2022). Liu et al. (2016) demonstrated that the application of controlled-release fertilizer not only decreases nitrogen runoff and leaching loss by 22% and 25%, respectively, but also augments a 9.8% increase of soil residual mineral nitrogen, together with a 5.5% increase in wheat production compared to traditional fertilizers. Numerous studies have revealed that the decreasing size of NMs benefits the increase of the surface mass ratio of particles. Various nutrient ions can be desorbed and adsorbed steadily and slowly for an extended period, significantly improving the uptake process in crops by adjusting the demanded release.

## 2.7 Challenges and Considerations of Nanofertilizers

1. The rapid expansion of nanofertilizers in agriculture presents promising advancements in enhancing nutrient availability, reducing losses, and promoting plant growth. However, these specialized fertilizers also raise significant environmental and health concerns that must be carefully assessed and managed to ensure their responsible and sustainable deployment. Key challenges and considerations include: Leaching of toxic nanoparticles from nanofertilizers poses risks to humans, animals, and aquatic life. Metal-based nanoparticles are less toxic compared to their oxides, but the irreversible toxicity in freshwater bodies and air raises significant concerns.
2. Nanofertilizers can affect non-target organisms, such as soil microbes, which plays a crucial role in nutrient cycling and aquatic life. Toxicity varies based on composition, size, and surface properties of the nanoparticles.
3. Potential entry of nanoparticles into the food chain, raising questions about bioaccumulation and long-term effects on ecosystems and human health.
4. Limited information on the toxicological effects of nanoparticles on human health, emphasizing the need for further research.
5. Risks of exposure to nanoparticles (particularly airborne) for individuals involved in nanofertilizer production and application, necessitating proper safety measures.
6. Concerns about skin exposure to nanofertilizers for workers during handling and application, requiring protective measures.
7. Complex manufacturing processes may increase production costs, limiting availability and hindering widespread adoption of nanofertilizers.
8. The concentration and effects of nanoparticles on distinct crop species is still at a nascent stage and needs further study.

9. Toxic nanoparticles can disrupt respiration and photosynthesis. Nanoparticles, particularly those made of oxides like $TiO_2$, can remain in the soil for an extended period and exhibit toxicity to plants.
10. A detailed understanding the potential differences between plants in nutrient transformation and assimilation via nanoparticles and traditional fertilizer material is required.
11. Effective outcomes often necessitate repetitive application of nanomaterials due to factors such as low minimum effective concentration, leaching, photolysis, and hydrolysis.

Addressing these challenges is essential for integrating nanofertilizers into sustainable agricultural practices. Despite these concerns, the potential benefits of nanofertilizers in meeting future food demands provide a promising solution for agriculture, as long as environmental impacts and health risks are carefully managed and mitigated.

## 2.8 Future Directions and Research Needs

The development of nanofertilizers has made significant strides in enhancing plant growth, crop improvement, and fertilization. However, several critical areas require further research and development to ensure their effective and sustainable use. Key future directions include improving fabrication processes to be cost-effective, understanding bio-performance, and elucidating uptake mechanisms and tolerable limits for different crops. Detailed studies on ecotoxicity are essential to mitigate potential environmental impacts. Challenges such as high processing costs, scalability, public perception, safety concerns, and regulatory standards need innovative solutions and robust global frameworks. Expanding research into plant physiology and developing real-time monitoring tools like nanobiosensors can enhance precision agriculture. Establishing the safety and efficacy of nanofertilizers through long-term studies is crucial for food security in a changing climate. Field-scale demonstrations are necessary to validate benefits in real-world settings, and using biologically sourced nanomaterials can address toxicity concerns, paving the way for sustainable and resilient agricultural practices.

## 2.9 Conclusion

In conclusion, the integration of nanofertilizers in agriculture presents a transformative potential to address pressing challenges in global food security, environmental sustainability, and agricultural practices. The extensive review underscores the multifaceted applications of nanofertilizers, particularly in enhancing crop yield and quality while mitigating environmental pollution. Nanofertilizers, with their controlled release mechanisms and improved nutrient efficiency, offer a sustainable alternative to conventional fertilization practices. Precision agriculture, supported by nanosensors, enables real-time monitoring and informed decision-making, optimizing resource utilization and promoting eco-friendly farming. Moreover, nanofertilizers showcase promise in enhancing water use efficiency, crucial in regions grappling with water scarcity, thereby contributing to sustainable water management. However, as nanofertilizer technology advances, regulatory and safety concerns emerge, necessitating comprehensive assessments of potential health and environmental risks associated with nanomaterials. Future prospects lie in continued research to understand the prolonged effects of nanoparticles, explore encapsulation techniques for nano agrochemicals, and address concerns related to toxicity and safety. Responsible research, public awareness, and regulatory frameworks are imperative to ensure the safe and sustainable implementation of nanofertilizers in agriculture.

## References

Abbas M, Yan K, Li J, Zafar S, Hasnain Z, Aslam N, Iqbal N, Hussain SS, Usman M, Abbas M, Tahir M, Abbas S, Abbas SK, Qiulan H, Zhao X, El-Sappah AH (2022) Agri-nanotechnology and tree nanobionics: augmentation in crop yield, biosafety, and biomass accumulation. Front Bioeng Biotechnol 10:853045

Abdel-Aziz HMM, Hasaneen MNA, Omer AM (2016) Nano chitosan-NPK fertilizer enhances the growth and productivity of wheat plants grown in sandy soil. Span J Agric Res 14:17

Abou-Zeid HM, Ismail GSM, Abdel-Latif SA (2020) Influence of seed priming with zno nanoparticles on the salt-induced damages in wheat (Triticum aestivum L.) plants. J Plant Nutr 44(1):1–15

Adisa IO, Pullagurala VLR, Peralta-Videa JR, Dimkpa CO, Elmer WH, Gardea-Torresdey JL et al (2019) Recent advances in nano-enabled fertilizers and pesticides: a critical review of mechanisms of action. Environ Sci Nano 6:2002–2030

Al-Shaabani EM, AL-Obeid AK, Alrubaye AA, Wissam A, Ahmed YM, Al-maamri HA, Al-Budeiri MH (2023) Effect of surface, subsurface drip irrigation and nanotechnology perfusion methods on irrigation efficiency and saved water ratio in plowed and unplowed soils. IOP Conf Ser. Earth Environ Sci 1222:012028

Aneja VP, Schlesinger WH, Erisman JW (2009) Effects of agriculture upon the air quality and climate: research, policy, and regulations. Environ Sci Technol 43:4234–4240

Banger K, Yuan M, Wang J, Nafziger ED, Pittelkow CM (2017) A vision for incorporating environmental effects into nitrogen management decision support tools for U.S. Maize production. Front Plant Sci 8:1270

Borges R, Wypych F, Petit E, Forano C, Prevot V (2019) Potential sustainable slow-release fertilizers obtained by mechanochemical activation of MgAl and MgFe layered double hydroxides and $K_2HPO_4$. Nano 9:183

Das CK, Srivastava G, Dubey A, Roy M, Jain S, Sethy NK, Saxena M, Harke S, Sarkar S, Misra K (2016) Nano-iron pyrite seed dressing: a sustainable intervention to reduce fertilizer consumption in vegetable (beetroot, carrot), spice (fenugreek), fodder (alfalfa), and oilseed (mustard, sesamum) crops. Nanotechnol Environ Eng 2016(1):2

Davarpanah S, Tehranifar A, Davarynejad G, Abadía J, Khorassani R (2016) Effects of foliar applications of zinc and boron nano-fertilizers on pomegranate (Punica granatum cv. Ardestani) fruit yield and quality. Sci Hortic 210:57–64

Davidson EA (2009) The contribution of manure and fertilizer nitrogen to atmospheric nitrous oxide since 1860. Nat Geosci 2:156–157

Dekkers S, Krystek P, Peters RJ, Lankveld DX, Bokkers BG, van HoevenArentzen PH, Bouwmeester H, Oomen AG (2011) Presence and risks of nanosilica in food products. Nanotoxicology 5:393–405

DeRosa MC, Monreal C, Schnitzer M, Walsh R, Sultan Y (2010) Nanotechnology in fertilizers. Nat Nanotechnol 5:91

Dubey A, Mailapalli DR (2016) Nanofertilisers, nanopesticides, nanosensors of pest and nanotoxicity in agriculture. In: Lichtfouse E (ed) Sustainable agriculture reviews, vol 19. Springer, Cham, pp 307–330

Duhan JS, Kumar R, Kumar N, Kaur P, Nehra K, Duhan S (2017) Nanotechnology: the new perspective in precision agriculture. Biotechnol Rep 15:11–23

FAO (2018) The future of food and agriculture—alternative pathways to 2050. Summary version. Rome, Italy, 60 Licence: CC BY-NC-SA 3.0 IGO

Gaur M, Misra C, Yadav AB, Swaroop S, Maolmhuaidh FÓ, Bechelany M, Barhoum A (2021) Biomedical applications of carbon nanomaterials: fullerenes, quantum dots, nanotubes, nanofibers, and graphene. Materials 14:5978

Ghormade V, Deshpande MV, Paknikar KM (2011) Perspectives for nano-biotechnology enabled protection and nutrition of plants. Biotechnol Adv 29:792–803

Golbashy M, Sabahi H, Allahdadi I, Nazokdast H, Hosseini M (2016) Synthesis of highly intercalated urea-clay nanocomposite via domestic montmorillonite as eco-friendly slow-release fertilizer. Arch Agron Soil Sci 63:84

Hunter MC, Smith RG, Schipanski ME, Atwood LW, Mortensen DA (2017) Agriculture in 2050: recalibrating targets for sustainable intensification. Bioscience 67:386–391

Iqbal M, Umar S (2019) Nano-fertilization to enhance nutrient use efficiency and productivity of crop plants. In: Nanomaterials and Plant Potential. Springer, pp 473–505

Jakhar AM, Aziz I, Kaleri AR, Hasnain M, Haider G, Ma J, Abideen Z (2022) Nano-fertilizers: a sustainable technology for improving crop nutrition and food security. NanoImpact 2022(27):100411

Jyothi TV, Hebsur NS (2017) Effect of nanofertilizers on growth and yield of selected cereals - a review. Agric Rev 38:112–120

Kah M, Kookana RS, Gogos A, Bucheli TD (2018) A critical evaluation of nanopesticides and nanofertilizers against their conventional analogues. Nat Nanotechnol 13:677–684

Kale AP, Gawade SN (2016) Studies on nanoparticle induced nutrient use efficiency of fertilizer and crop productivity. Green Chem 2:88–92

Kale RD, Kane PB (2017) Colour removal using nanoparticles. Text Cloth Sustain 2(1):1–7

Kamiab F, Zamanibahramabadi E (2016) The effect of foliar application of nano-chelate super plus ZFM on fruit set and some quantitative and qualitative traits of almond commercial cultivars. J Nuts 7:9–20

Kole C, Kole P, Randunu KM, Choudhary P, Podila R, Ke PC, Rao AM, Marcus RK (2013) Nanobiotechnology can boost crop production and quality: first evidence from increased plant biomass, fruit yield and phytomedicine content in bitter melon (Momordica charantia). BMC Biotechnol 13:37

Kottegoda N, Munaweera I, Madusanka N, Karunaratne V (2011) A green slow-release fertilizer composition based on urea-modified hydroxyapatite nanoparticles encapsulated wood. Curr Sci 1:73 78

Lee Y-C, Moon J-Y (2020) Introduction to nanotechnology and bionanotechnology. In: Lee Y-C, Moon J-Y (eds) Introduction to bionanotechnology. Springer, Singapore, pp 1–14

Liu R, Lal R (2015) Potentials of engineered nanoparticles as fertilizers for increasing agronomic productions. Sci Total Environ 2015(514):131–139

Liu J, Zhang Y, Zhang Z (2009) The application research of nano-biotechnology to promote increasing of vegetable production. Hubei Agric Sci 1:041

Liu RH, Kang YH, Pei L, Wan SQ, Liu SP, Liu SH (2016) Use of a new controlled loss-fertilizer to reduce nitrogen losses during winter wheat cultivation in the Danjiangkou reservoir area of China. Commun Soil Sci Plant Anal 47:1137–1147

López-Vargas ER, Ortega-Ortíz H, Cadenas-Pliego G, de Alba RK, Cabrera de la Fuente M, Benavides-Mendoza A, Juárez-Maldonado A (2018) Foliar application of copper nanoparticles increases the fruit quality and the content of bioactive compounds in tomatoes. Appl Sci 8(7):1–15

Mali SC, Raj S, Trivedi R (2020) Nanotechnology a novel approach to enhance crop productivity. Biochem Biophys Rep 24:100821

Manikandan A, Subramanian KS (2017) Study on mitigation of ammonia volatilization loss in urea through adsorbents. J Appl Nat Sci 9:688–692

Manjunatha SB, Biradar D, Aladakatti Y (2016) Nanotechnology and its applications in agriculture: a review. J Farm Sci 29:1–13

Maroufpoor N, Mousavi M, Hatami M, Rasoulnia A, Lajayer BA (2019) Mechanisms involved in stimulatory and toxicity effects of nanomaterials on seed germination and early seedling growth. In: Ghorbanpour M, Wani SH (eds) Advances in phytonanotechnology. Academic Press, Cambridge, pp 153–181

Mikkelsen R (2018) Nanofertilizer and nanotechnology: a quick look. Better Crops 102:18–19

Milani S, Baldelli Bombelli F, Pitek AS, Dawson KA, Rädler J (2012) Reversible versus irreversible binding of transferrin to polystyrene nanoparticles: soft and hard Corona. ACS Nano 6(3):2532–2541

Millan G, Agosto F, Vazquez M, Botto L, Lombardi L, Juan L (2008) Use of clinoptilolite as a carrier for nitrogen fertilizers in soils of the Pampean regions of Argentina. Cienc Investig Agrar 35:293–302

Mirzajani F, Askari H, Hamzelou S, Farzaneh M, Ghassempour A (2013) Effect of silver nanoparticles on Oryza sativa L. and its rhizosphere bacteria. Ecotoxicol Environ Saf 88:48–54

Monreal C, DeRosa M, Mallubhotla S, Bindraban P, Dimkpa C (2016) Nanotechnologies for increasing the crop use efficiency of fertilizer-micronutrients. Biol Fertil Soils 52:423–437

Montalvo D, McLaughlin MJ, Degryse F (2015) Efficacy of hydroxyapatite nanoparticles as phosphorus fertilizer in andisols and oxisols. Soil Sci Soc Am J 79(2):551–558

Parveen A, Mazhari BBZ, Rao S (2016) Impact of bio-nanogold on seed germination and seedling growth in Pennisetum glaucum. Enzym Microb Technol 2016(95):107–111

Raimondi G, Maucieri C, Toffanin A, Renella G, Borin M (2021) Smart fertilizers: what should we mean and where should we go? Ital J Agron 2021:16

Raliya R, Saharan V, Dimkpa C, Biswas P (2018) Nanofertilizer for precision and sustainable agriculture: current state and future perspectives. J Agric Food Chem 66:6487–6503

Rico CM, Majumdar S, Duarte-Gardea M, Peralta-Videa JR, Gardea-Torresdey JL (2011) Interaction of nanoparticles with edible plants and their possible implications in the food chain. J Agric Food Chem 59:3485–3498

Sabir A, Yazar K, Sabir F, Kara Z, Yazsci MA, Goksu N (2014a) Vine growth, yield, berry quality attributes and leaf nutrient content of grapevines as influenced by seaweed extract (Ascophyllum nodosum) and nanosize fertilizer pulverizations. Sci Hort 175:1–8

Sabir S, Arshad M, Chaudhari SK (2014b) Zinc oxide nanoparticles for revolutionizing agriculture: synthesis and applications. Scientific World J 2014:925494

Saied HHM (2018) Response of Keitte mango trees to spraying nano NPK Mg fertilizers. Researcher 10:1–5

Sarkar S, Datta SC, Biswas DR (2014) Effect of fertilizer loaded nanoclay/superabsorbent polymer composites on nitrogen and phosphorus release in soil. Proc Natl Acad Sci 85:415

Sekhon BS (2014) Nanotechnology in Agri-food production: an overview. Nanotechnol Sci Appl 7(2014):31

Servin AD, Michel HC, Viezcas JAH, Diaz BC, Videa JRP, Torresdey JLG (2012) Synchrotron micro-XRF and micro-XANES confirmation of the uptake and translocation of TiO2 nanoparticles in cucumber (Cucumis sativus) plants. Environ Sci Technol 46:7637–7643

Sharif R, Mujtaba M, Ur Rahman M, Shalmani A, Ahmad H, Anwar T et al (2018) The multifunctional role of chitosan in horticultural crops; a review. Molecules 23:872
Shojaei TR, Salleh MAM, Tabatabaei M, Mobli H, Aghbashlo M, Rashid SA, Tan T (2019) Applications of nanotechnology and carbon nanoparticles in agriculture. In: Suraya AR, Raja NIRO, Mohd ZH (eds) Synthesis, technology and applications of carbon nanomaterials. Elsevier, Amsterdam, pp 247–277
Singh SK, Pathak R, Kalia RK (2019) Perspectives of nanomaterial in sustainable agriculture, nanoscale engineering in agricultural management. CRC Press, pp 64–80
Song HW, Yuan WM, Jin P, Wang W, Wang XF, Yang LM, Zhang YF (2016) Effects of chitosan/nano-silica on postharvest quality and antioxidant capacity of loquat fruit during cold storage. Post Biol Tech 119:41–48
Suppan S (2013) Nanomaterials in soil. Institute for Agriculture and Trade Policy
Tan W, Du W, Barrios AC, Armendariz R, Zuverza-Mena N, Ji Z, Chang CH, Zink JI, Hernandez-Viezcas JA, Peralta-Videa JR, Gardea-Torresdey JL (2017) Surface coating changes the physiological and biochemical impacts of nano-TiO2 in basil (Ocimum basilicum) plants. Environ Pollut 222:64–72
Tarafder C, Daizy M, Alam MM, Ali MR, Islam MJ, Islam R et al (2020) Formulation of a hybrid nanofertilizer for slow and sustainable release of micronutrients. Acs Omega 5:23960–23966
Wang Y, Lin Y, Xu Y, Yin Y, Guo H, Du W (2019) Divergence in response of lettuce (var. ramosa Hort.) to copper oxide nanoparticles/microparticles as potential agricultural fertilizer. Environ Pollut Bioavail 31(1):80–84
Wilson MA, Tran NH, Milev AS, Kannangara GSK, Volk H, Lu GQM (2008) Nanomaterials in soils. Geoderma 146(1–2):291–302
Yadav A, Yadav K, Abd-Elsalam KA (2023) Nanofertilizers: types, delivery and advantages in agricultural sustainability. Agrochemicals 2:296–336
Yata VK, Tiwari BC, Ahmad I (2017) Research trends and patents in nano-food and agriculture. In: Ranjan S, Dasgupta N, Lichtfouse E (eds) Nanoscience in food and agriculture 5. Springer International Publishing, Cham, pp 1–20
Zahedi SM, Hosseini MS, Meybodi NDH, Teixeira da Silva JA (2019) Foliar application of selenium and nano-selenium affects pomegranate (Punica granatum cv. Malase Saveh) fruit yield and quality. S Afr J Bot 124:350–358

# Chapter 3
# Transformations of Nanofertilizers in Soil and Their Uptake by Plants

**Tapan Adhikari**

**Abstract** Nanofertilizers, designed to enhance nutrient delivery and uptake in plants, have garnered increasing attention for their potential to revolutionize agricultural practices. This chapter provides a comprehensive overview of the transformations of nanofertilizers in soil and their subsequent uptake by plants, focusing on chemical, physical, and biological processes. Chemical transformations involve the dissolution of nanofertilizers and subsequent release of nutrient ions, influenced by factors such as pH, soil composition, and organic matter content. Physical transformations encompass changes in nanoparticle size, aggregation, and surface properties, affecting their mobility and availability to plants. Biological transformations mediated by soil microorganisms play a crucial role in nanoparticle stability, biotransformation, and plant-microbe interactions. Understanding the mechanisms of nanofertilizer uptake by plants is essential for optimizing their efficacy and minimizing environmental risks. Nanofertilizers can be taken up by plants through both foliar and root pathways. Foliar uptake involves the direct absorption of nanoparticles through leaf stomata or cuticles, while root uptake occurs through interactions with the rhizosphere, including adsorption, diffusion, and endocytosis processes. Moreover, the chapter discusses the factors influencing nanofertilizer uptake by plants, such as nanoparticle properties (size, shape, surface charge), plant species, growth stage, and environmental conditions. Insights into the fate and behavior of nanofertilizers in soil-plant systems are crucial for sustainable agriculture, ensuring efficient nutrient use, minimizing environmental impacts, and enhancing crop productivity. This chapter integrates current research findings and highlights future research directions to advance our understanding of nanofertilizer transformations and their implications for agricultural sustainability.

**Keywords** Transformations · Nanofertilizers · Mechanism · Soil · Plant · Microbes · Rhizosphere

---

T. Adhikari (✉)
ICAR-Indian Institute of Soil Science, Bhopal, Madhya Pradesh, India



A. K. Patra (ed.), *Nanofertilizers in Agriculture*, Sustainability Sciences in Asia and Africa, https://doi.org/10.1007/978-981-97-8112-6_3

## 3.1 Introduction

Currently, nanofertilizer is a groundbreaking, effective, and nature friendly substitute for chemical fertilizer. The nutrient release pattern from nanofertilizers sustenance crop growth and soil ecosystem. The conversion of nanofertilizer in soil and plant system is to comprehend the safe delivery of nanofertilizers in the agricultural field for production of crops (Lowry et al. 2012). The various transformations of nanofertilizer, viz. physical, chemical, and biological, may happen in the soil medium, due to its highly dynamic and reactive nature (Valsami-Jones and Lynch 2015). During these processes, the original profile of nanofertilizer is altered to new properties, which subsequently affect its behavior, like fate in soil solution and translocation in the plant system. For instance, some metal-based nanofertilizers, such as silver (Ag) (Das et al. 2018), copper (Cu) (Conway et al. 2015), and zinc oxide (ZnO) (David et al. 2012), may partially dissolve in the soil matrix. Some naturally occurring coating materials like organic molecules, root exudates, etc., may interact with these nanofertilizers and as a result accelerates or reduces the dissolution rate (Kumar et al. 2018). The ions released from nanofertilizer might be taken up by the plant in the form of complexes with other constituents from soil solution or directly. The chemical structures like silicon dioxide ($SiO_2$), titanium dioxide ($TiO_2$), etc. of some specific nanofertilizers are insistent in soil system. Certain physical transformations including aggregation, agglomeration, or natural coating formation may occur, which eventually change the basic chemistry and surface charge of the nano product and subsequently influence its bioavailability and nature. Generally, specific transformation processes change the function of nanofertilizers, which may arise in the soil matrix without interacting with the plant or at the interface of plant root and nanoparticle in the rhizosphere zone. Still the understanding of the transformation of nanofertilizers is under investigation in soil and, particularly, in plant systems. The chapter thus concentrates on mechanisms that deal with nanofertilizer conversion and various processes in the soil matrix and relates these intricate processes with the uptake, movement, and assimilation of nanofertilizers in plant systems.

## 3.2 Transformations of Nanofertilizers in Soil

The literal meaning of transformation is change in form, appearance, nature, or character. Because of their tiny size, nanomaterials (NMs) possess a large number of atoms on their surface, which creates unique features and higher reaction rate in comparison to bulk material of similar atomic conformation. Nanofertilizers are physiochemically dynamic and highly reactive in soil matrix. Several transformations including redox reactions, dissolution, reactions with biomacromolecules, and aggregation may happen in soil. These transformations will change the behavior and movement of nanofertilizers. The character and ambit of these conversions should be comprehended prior to delivery of the NMs in the agricultural field as a fertilizer. For instance, silver nano particles undergo redox reaction and may become sulfidized in soil solution. By the process of sulfidation, the aggregation state, surface chemistry, charge, and subsequent release rate of $Ag^+$ ions will change (Gogos et al. 2018) and their existence in soil solution may also alter. Likewise, equivalent to protein coronas (David et al. 2012), nanoscale coating of complexed organic substances on NMs alters their aggregation and precipitation properties. Because soil system is highly dynamic and random in nature, the physicochemical changes and subsequent reactions occur in the soil matrix, significantly intricating the understanding of the fate of nanofertilizers in soil. Still the nature and extent of conversions predicted for nano particles in soil system is under investigation and precise knowledge is to be documented. Nanofertilizers transformations are the most important processes governing its interactions with soil constituents. Various transformations like biologically mediated transformations including interactions with macromolecules, biomacromolecules, as well as physical and chemical.

### *3.2.1 Physical Transformations*

Uncoated or coated nanofertilizers once discharged into the soil environment interact with different soil constituents and, depending on the conditions, undergo different physical phenomena like aggregation, agglomeration, physisorption, etc. Agglomeration of NPs reduces the decreases in specific surface areas and subsequently reduces its reactivity in soil solution. The enhancement of aggregate size influences their transport in porous media, precipitation, fate, and bioavailability. NP aggregation turning into clusters over time is expected, except in the case of incidental coatings which prevent aggregation. Aggregation may be categorized into two forms like homoaggregation, which occurs among similar nano particles, or aggregation process between a nano particle and other dominant soil constituents. Homoaggregation is a combination of nano particles of similar character activated by the compressed electrical double layers on the nano particle surfaces under high ionic strength conditions. Longer combination processes cause bigger sizes of particles and bunch form. Generally, the higher concentration of specific soil constituents compared to NMs cause the formation of heteroaggregates. In most cases, due to occurrence of aggregation, the suspended nano particle number reduces, with subsequent rise in sizes. For instance, zero valent iron of 30–70 nm diameter rapidly agglomerated in solution and converted into bigger aggregates, significantly reducing their movement. When comparatively larger particles like clay bind NMs, this results in heteroaggregation, and subsequently changes the fate of NM. The NM–clay heteroaggregates of nano particles and clay ultimately exist as clay particles rather than NMs. Aggregation reduces the original surface area of the nano particles, thus reducing their reactivity in soil matrix.

### 3.2.2 Chemical Transformations

Chemical transformation infers constitutional or configuration changes of the nanofertilizers. Common chemical transformation of metal-based nanofertilizers is termed as dissolution, and this alteration procedure is usually convoyed by other chemical reactions including redox reactions, chelation, sulfidation, phosphorylation, etc. Redox reactions are conjugated methods in soil solution and encompass electron transmission within chemical components. Several nanofertilizers are comprised of or possess components that take part in reduction and oxidation reactions or together simultaneously in soil solution. For example, NMs like silver (Kumar et al. 2018; Gogos et al. 2018) and Fe (Banerjee and Jain 2018). Nano particles of Ceria may possess both Ce(IV) and Ce(III) and subsequently after sorption with bigger molecules, which in turn cause alteration in the ratio of Ce(III)/Ce(IV) on the surface of NP (Kampe et al. 2018). Ag-nanofertilizers release $Ag^+$ ion by oxidation process (Singh et al. 2019). During the process of oxidation, an oxide layer gets adsorbed on the surface of nanofertilizers that ultimately reduce another oxidation and rates of dissolution. But removal of the oxidation layer may enhance the release of $Ag^+$ ions. Furthermore, under transformation progression, oxidation reduction reaction and suspension, like complexation, might happen for metal-based nanofertilizers, such as iron (Kumar et al. 2018), copper (Gogos et al. 2018), zinc (Banerjee and Jain 2018), and silver (Kampe et al. 2018), sulfidation or silver chlorination (Singh et al. 2019), or cerium (Zhang et al. 2017a) and zinc (Xiao et al. 2018) phosphorylation. Moreover metal-based nanofertilizers as well as carbon-based nanomaterials can also transform through oxidation and reduction reactions. Aromatic hydrocarbons (PAH) may be produced by oxidation phenomena through fractional degradation and subsequently release $CO_2$. The respiration activities of bacteria lead to reduction of graphene oxide in soil. On the contrary to oxidizing environments prevailing in upland soils, reduction phenomena dominate in carbon-rich sediments and cause reduction of nano particles. Decomposition of coating materials like gum arabic on silver NPs occur due to sunlight exposure and cause aggregation and precipitation from suspension. Other nanofertilizers may be reduced or oxidized by sunlight, resulting in change in redox state and charge. The surface characteristics of nano materials and their residence time in suspension are controlled by sulfidation and dissolution phenomena. This may occur for NMs synthesized from Class B soft metal cations, since they form partially soluble metal oxides and, moreover, of their strong attraction for ligands like inorganic and organic sulfide. Sparingly soluble metal sulfide coating on surface of nano particles causes modification of the surface charge and as a result aggregation occurs. Dissolution and/or sulfidation rates of nano particle in the soil matrix are primarily based on their size, shape, capping material, etc. The fate of nano materials in soil solution is generally governed by sorbed organic and inorganic ligands or larger-molecules on their surfaces. On the contrary, polymer coatings enhance the movement of nano particles in soil system, which is due to reduction of their attachment to silica surfaces. Thiol groups containing organic ligands may cause NM dissolution, change in charge, and stability against aggregation.

### 3.2.3 Biologically Mediated Transformations

Biotransformation is facilitated by biological processes. Nanofertilizers may interact with the root surface in the rhizospheric zone, or aggregate inside plant systems in xylem and phloem sap. Different phenomena, including redox reactions, chelation, dissolution, or degradation, occur during the transport of nanofertilizers through plant cell wall or cell membrane, cytoplasm, vacuole. Different geochemical and biochemical processes including carbon and nitrogen cycles occur in soil and are mainly governed by bacteria (Ren et al. 2018). During this process, through different biological activities, like respiration, the root exudates secretion (Salas et al. 2010; Guo et al. 2017) and can activate several chemical transformations of nanofertilizers. Basic properties and specifically the size of nanofertilizers modified due to interaction with plant metabolites exist at leaf and root surface (Zhang et al. 2012a). The microbial community at rhizosphere controls the transformation of nanofertilizers. Biologically mediated transformations occur at plant–soil interface, which affect the fate and behavior of coated and naked nanofertilizers, including size, charge, aggregation, agglomeration state, and reactivity, and control the mobility and bio-availability of nano nutrients. Moreover, stabilizer or organic capping agent and small anion or polymer are generally used to coat the nanofertilizer to increase its efficacy. The crystalline phase of ZnS nanoparticles may change due to coordination with water molecules (Singh et al. 2019). Bacteria can alter the biodegradable capping agent. Certainly, the rate, extent, and type of conversions of nanonutrient are governed by its initial physicochemical properties and surface conditions. Prominently, several alterations are not so rapid and not reversible in nature and hence cannot be explained by the thermodynamics principle. Once released into the soil matrix, uncoated or coated nanofertilizers are exposed to changes through reaction with organic matter, which substantially alter the properties and behaviors of the nanofertilizers. The unique properties and influences of nanofertilizers are primarily based on the amount of protein coronas and soil components adsorbed on their surfaces. During interaction, the consistency and viscosity of humic substances were governed by the particle physicochemical properties and the soil matrix. The effects of NOM on nano nutrients are intricate and very difficult to predict which may change the charge and steric stabilization of nanofertilizers, in certain cases it results in bridging flocculation. Moreover, dispersion of nano particles also occurs in the presence of humic substances. Homoaggregation and heteroaggregation may also occur in case of humic substance coatings on nanofertilizers. The dissolution rate of nanofertilizers was minimized as a result of coating of organic matter on its surface.

### 3.2.4 Soil Properties

Soil is a heterogeneous mixture of different primary and secondary minerals, inorganic and organic substances and air and water filled pore spaces. The different soil parameters including texture, pH, ionic concentration, oxidation and reduction potential, moisture content, and organic carbon content collectively control the transformation of nanofertilizers in soil matrix. Transformations of nano nutrients occur in soil solution prior to interaction with the rhizosphere zone. The existence of different metal ions, minerals, and other inorganic matter (Baalousha 2017) and organic compounds like root exudates, fulvic acid, and humic acids (Duan et al. 2017) causes physical transformation of nanofertilizers. The rate of aggregation of nanofertilizers varies depending on the soil texture. Soil texture consists of different size specific solid materials including gravel, sand, silt, and clay (Jahn et al. 2006). The existence of negative sites on clayey and silty soils favors the sorption of cations (Manrique et al. 1991). Hence, positively charged nanofertilizers adsorbed strongly on those soils and more heteroaggregation occurs as compared to sandy soil. Soil texture also plays pivotal roles in influencing the soil water content and redox potential by controlling the air and water content (Adamu and Aliyu 2012). As clayey soil possesses higher water holding capacity and more micro pore space as compared to sandy soil, anaerobic conditions prevail in soils with high clay content. Hence, the reduction of nanofertilizers is prevailing more in clayey soil than in sandy soil (Li et al. 2017).

Conversely higher oxygen content prevails in sandy soils which could favor oxidative changes of nanofertilizers specifically for iron and copper nanofertilizers. Natural organic matters (NOMs) comprise ≈1–5% of agricultural soil composition (Brady and Weil 2008). It consists of plant derived protein, humic substances, and polysaccharides etc. Organic material coating or "corona" on the surface of nanofertilizers modifies its fate and reactivity in soil system. Relatively higher availability of NOMs in soil cause substantial and dynamic transformation of nanofertilizers in soil matrix. Nevertheless, the composition and structure of the organic substances and specific soil conditions, primarily pH and ionic strength, determine the impacts of this alteration on the physicochemical properties of nanofertilizers (Lin et al. 2017). Categorically NOMs are of two types: hydrophilic and hydrophobic. Humic matter, the hydrophilic portion represents 60–80% of NOMs (Schnitzer and Khan 1975).

Adsorption of hydrophilic and hydrophobic fractions to nanofertilizer surfaces creates opposite impacts in terms of surface charge. Larger molecular weight NOMs develop coherent and thick coatings, which partly or entirely cover the surface of nanofertilizers. Surface coating with hydrophilic and hydrophobic fractions of nanofertilizers influences the charge on the surface and in case of larger particles of organic matter totally cover the surface area of nanofertilizers. The humic matter properties govern the electrostatic stabilization of nanofertilizers (Degenkolb et al. 2019) as well as flocculation and aggregation because of linking effects amongst NOM molecules (Luo et al. 2018). The ionic strength determines the stability of nanofertilizers in soil suspension. According to Derjaguin, Landau, Verwey, and Overbeek (DLVO) theory, the stability of colloidal particles is regulated by the comparative equilibrium between the electrostatic repulsion and the van der Waals attraction (Ohshima 2012). Nanofertilizers suspension is unchanging as an adequate charge exists in the electrostatic double layer which subsidizes to the repulsion. Noteworthy higher ionic strength compresses the electrostatic double layer and causes agglomeration of nanofertilizers (Zhang et al. 2012b). Soil pH, valency, and type of ions control the impact on nano particles suspension. In general, lower valent ions have lesser effects than higher valent ions (Tang and Cheng 2018). For instance, at same ionic strength, the $Ca^{2+}$ electrolyte results in larger agglomeration of negatively charged citrate functionalized Ag nanofertilizers as compared to $Na^{+}$ (Badawy et al. 2010). $Ca^{2+}$ ions interact with the COOH groups on citrate that decrease the charge of silver nanofertilizers as well as brings succeeding agglomeration. On the other hand, zeta potential is not influenced by $Na^{+}$ or $Ca^{2+}$ which cause agglomeration of neutral PVP-coated Ag nanofertilizers irrespective of pH range. Hence occurrence of ions may not induce nanofertilizers stability in soil solution (Badawy et al. 2010). Soil pH plays a pivotal role in nanofertilizers transformation. For instance, it affects agglomeration as various nanofertilizers have different isoelectric points (IEP). The IEP differs markedly amongst the types of nanofertilizers, varying from low pH to high pH value. Hence, the potential impact of pH on the stability of nanofertilizers in soil is fairly dissimilar.

Soil pH influences the chemical transformation of nanofertilizers. Moreover, it is transformed into zincate ions, $Zn(OH)n^{(n-2)-}$, under higher pH conditions (pH > 12) (Tang et al. 2002). Redox reactions dominate in numerous crucial biogeochemical processes in soil and influence the conversion of nanofertilizers which are inclined to the oxidation reduction potential (Eh). The Eh value generally estimates the soil reducing or oxidizing capacity. The Eh governs the relative quantity of oxidants and reductants in soil solution (Bartlett and James 1993). For example, graphene oxide is reduced by the ferrous ions ($Fe^{2+}$) (Wang et al. 2015). Occurrence of nanofertilizers oxidation is more common in aerobic conditions, which subsequently form an oxide layer as a protective coating on the surface of some metallic nanofertilizers. On the other hand, nanofertilizers like Ag NPs, development of an $Ag_2O$ layer induces the $Ag^{+}$ ion release.

Anaerobic conditions favor the reduction of nanofertilizers. Experimental results revealed that in case of aerobic conditions under paddy, after 2 days' incubation 90% of Ag remained as Ag NPs in soil, while under anoxic conditions, 95% was reduced to $Ag_2S$ (Li et al. 2017). Furthermore, the soil microenvironment may sometimes play different roles. For instance, it was observed that the presence of extremely localized oxygen-enriched sites in a paddy soil can influence the transformation of nanofertilizers. This type of intricate condition may occur in reality in situ soil system and it may be considered during interpretation of nanofertilizers transformation. Chemical transformation of metal-based nanofertilizers in soil is mainly mediated by inorganic ligands like phosphate, chloride, and sulfide. Generally, sulfide anion is abundant in soil solution, and as per the Pearson acid-base concept sulfidation reaction occurs for nanofertilizers like Cu, ZnO, etc. (Pearson 1963).

Metal oxides are usually more soluble than their sulfide forms. Despite this, a weak sulfidation reaction produces low impact on zinc ion release from ZnO, while a high degree of sulfidation potentially checks the release kinetics of $Zn^{2+}$ (Ma et al. 2013). Soil pH determines the dominant species of Ag NPs in suspension. Generally, normal to higher pH conditions are conducive for Ag-sulfidation and the subsequent product $Ag_2S$ is stable chemically, whereas AgCl formation occurs at low pH conditions. Soil chemical composition governs the transformation of nanofertilizers. For instance, the soil microbes in particular bacteria reduce $CeO_2$ nanofertilizers to release $Ce^{3+}$ in soil and subsequently convert them into $CePO_4$. Again, biogenic organic acids release $Ce^{3+}$ by redissolution of $CePO_4$ which oxidized to $Ce^{4+}$ and converted into nanoscale $CeO_2$. Finally, the $CeO_2$ NPs accumulate on $CePO_4$ surface (Cervini-Silva et al. 2005). Conversion of metal ions to metallic nanoparticles occurs rightly in the plant system or by plant parts extracts (Gardea-Torresdey et al. 2002; Roy et al. 2017). The microbial, particularly, bacterial biomass plays a pivotal role in transformation of nanofertilizers in the soil solution. Bacteria-mediated physical and chemical transformation processes significantly affect the conversion of nanofertilizers. For instance, *Bacillus subtilis* reduce $CeO_2$ nanofertilizers at the surface and convert into $Ce^{3+}$ ions that could be complexed by carboxyl group present on bacterial cell wall (Xie et al. 2019). Oxidative degradation of the nanomaterial like carbon nanotubes occurs by the enzymes or radicals produced by bacteria (Allen et al. 2009). Similarly, nanofertilizers like graphene oxide can be reduced by facultative anaerobes *Shewanella*, which donate electrons during their respiration process (Salas et al. 2010). These reactions may occur at the micobial cell wall and inside the cell. The various microbially produced enzymes not only catalyze the different biogeochemical reactions, like decomposition of biological remains and essential plant nutrient cycling, but also mediate the transformation of nanofertilizers (Lewis et al. 2019). The varied and non-homogeneous nature of the soil makes the transformation process of nanofertilizers in soil complex in nature. In soil, various processes or phenomena arise concurrently depending on the existing soil environmental conditions including pH and oxygen content, which decide the dominant and final forms of nanofertilizers as well as their residence time in the soil.

## 3.3 Plant Uptake of Nanofertilizers

Generally, the unique characters of nanoparticle and different plant parameters control the uptake, translocation, and accretion of nanofertilizers in plant systems (Solanki et al. 2016). Ideally the nutrients should be applied to plants based on its demand, and uptake of nutrients is mainly driven by dissolution and ionic exchange phenomena. Nutrient absorption process is slow and nutrient supply to plant depends on nutrient concentration nearby rhizosphere zone. As zeolite possesses cation and anion exchange property, nano-zeolite releases K slowly and progressively for plant growth (Zhou and Huang 2007). Delivery of fertilizers to the plant through nano zeolite slows the release of nutrients and subsequent absorption into the plant by providing required quantity of nutrients as and when it is needed. The huge surface area of nano enhanced fertilizers allows to load a quantum of major essential nutrients, which may be taken up by plant based on the demand.

### *3.3.1 Leaf*

Generally physiological demands of plants govern the uptake of different nano-nutrients, which are absorbed by stomata, stigma, hydathodes, and trichomes that exist on leaf surface and subsequently passage through phloem and xylem (Rui et al. 2015). Nanoparticles enter into many organelles in the cytoplasm and influence different plant metabolic processes (Solanki et al. 2016). Apoplastic and symplastic pathways are usually the two routes for the movement of nano-nutrients inside the plant cell. In case of apoplastic pathway, the translocation of nano nutrients arises through the primary cell wall, the middle lamella, and the secondary cell wall and other spaces between parenchyma cells and the middle lamella. Moreover, in this process, the entry of bigger molecules is restricted based on size fraction (5–20 nm) segregation boundaries of plant cell walls. If the size of the pores (5–40 nm) is larger than the particle size of nano nutrients, they go inside plant cells directly through sieve-like structures of the cell walls. Plasmodesmata inside the plasma membrane helps the transport of nano nutrients from one cell to another cell in case of symplastic pathway. The nano nutrients translocate amongst the cells through the cell wall by endocytosis. The transport of the nano nutrients through the plant cell wall is regulated by the stomatal pore size, which varies from 5 nm to 40 nm (Devi et al. 2015), or by the trichrome base, and subsequently enters into the tissues. Movement based on the symplast path rest on the size exclusion principles of the plasmodesmata. The Casparian strip prevents the transport of nano nutrients into the vascular system (Corbo et al. 2016). Sometimes with the help of enzymatic reaction, larger size nano nutrients (>50 nm) may enter inside the cell wall. Based on agar culture experiment with CuO nanoparticles, it was established that they enter the maize plant cell membrane and accumulate in the cell vacuole (Adhikari et al. 2016). Similarly, foliar application of $CeO_2$ and Ag nanoparticles enhances their absorption by plant and subsequent translocation into the plant tissues of cucumber and lettuce respectively (Bartlett and James 1993; Lynch et al. 2014).

Entry of nano-nutrients into the plant system and the mechanisms leading to its absorption still raise several questions. Generally, nano nutrients may be applied to plants through different modes like soil application and foliar spray (Fig. 3.1). Nano-nutrients present in the soil solution interact with plant roots. No alien material can enter inside the plant cells. Plant cell wall hinders the entry of the elements which are not essential for its growth and acts as a barrier. The cell wall pore size generally ranges from 5 nm to 50 nm and controls the filtering properties. Henceforth, only nano nutrients or agglomerates with diameter less than the cell wall pore size can enter and deposit on the plasma membrane (Fig. 3.2). Another probable way of entry of nano nutrients into the plant system is that they can penetrate the cell walls

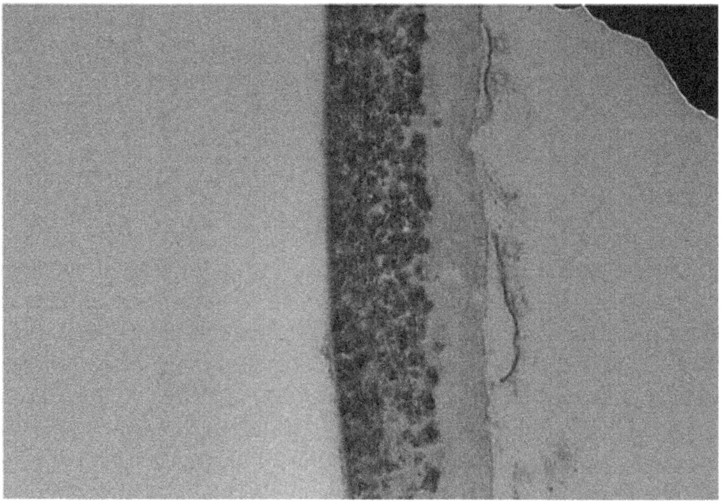

**Fig. 3.1** Deposition of CuO nanoparticle (<50 nm) in the outer wall of corn leaf epidermis

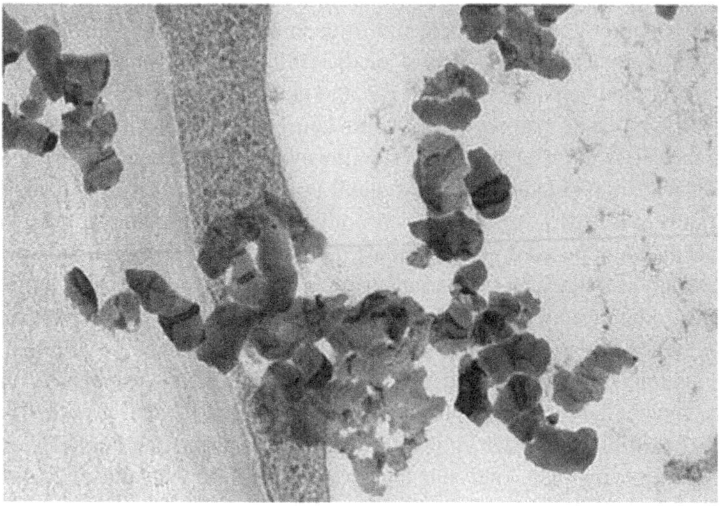

**Fig. 3.2** CuO nanoparticle (<50 nm) penetrating through the root cell wall of corn

by enlarging the pore size because of their tiny size and enormous kinetic energy. This process also helps to create larger pore sizes which enable uptake of nano nutrients. Furthermore, it was recorded that the activity of glucose-6-phosphate dehydrogenase enzyme was aggravated because of application of the nano nutrients (Fig. 3.3). It is assumed that the nano nutrients control the oxidative pentose phosphate pathway and increase the growth and yield of crops by enhancing the rate of

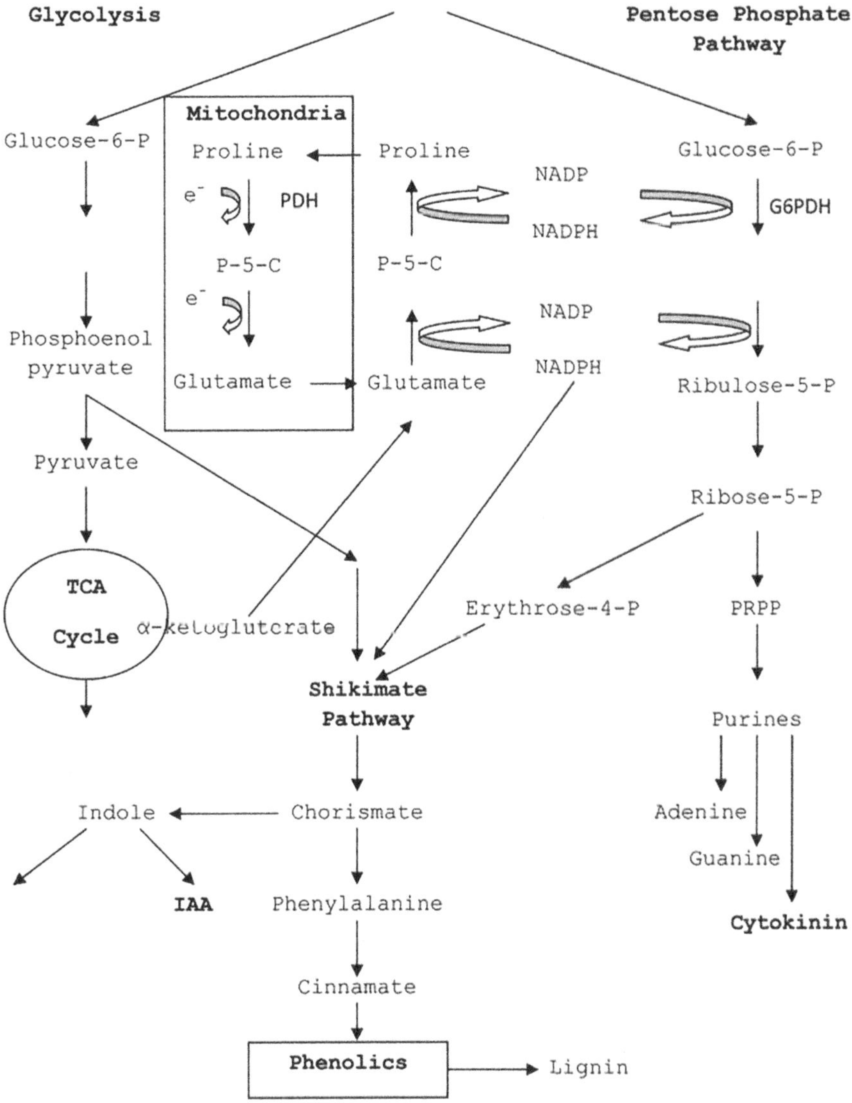


**Fig. 3.3** Phenolic compounds biosynthesis in plants (Shetty 2004)

photosynthesis. More intensive research work on this aspect is required to confirm whether nano nutrients within the plant system govern the yield parameters of crops.

## 3.3.2 Root

Nano nutrients enter inside the xylem by piercing the root epidermis and are translocated to the plant leaf. Generally, the size of nano nutrients varies from 2 nm to 20 nm, and they pass into the cell wall through pores (Kah et al. 2019). Root apex and lateral root points are the other routes of entry for nano nutrients into plant cells. Normally, the pore size of the cell walls varies from 3 nm to 8 nm (Chen et al. 2013), which restricts the entry of nano nutrients of larger sizes. But according to research findings, nano nutrient internalization inside the cells occurs through the creation of macro pores by nano nutrients in the cell walls (Zhang et al. 2019). For instance, the spherical size silica nano (14–200 nm) particles can be absorbed by *Arabidopsis thaliana* through the root (Aubert et al. 2012). Similarly, 40 nm Au particles were translocated from the roots into the shoots in *Solanum lycopersicum* (Koelmel et al. 2013). The nano nutrients enter into the plant cell through the root hairs. Consequently, encapsulated microspheres are dissolved by the root exudates, which contain organic acids or phenols and release the nano nutrients including Ca, Mg, Fe, Zn, etc. inside the cells (Schauer and Fernie 2006). Usually leaching loss of nutrients from conventional bulk fertilizers occur in the soil, which subsequently pollute the soil and water system. Besides that, some specific agro chemicals cause greenhouse gases and climate change. Controlled release of nano nutrients from nanofertilizers is one of the effective ways to enhance the efficiency of nanofertilizers.

For example, the release of chosen nutrients in protoplasts through mesoporous silica nanoparticles was reported by Torney et al. (Torney et al. 2007). In another investigation, double-layered hydroxide nanocomposites were used for release of nano nutrients in plant cells (Sharma et al. 2015). Reports revealed that integrated superabsorbent fertilizers based on a surface cross-linked process possessed slow-release property as well as super moisture retention capacity Wang et al. (2021). The transformation of nanofertilizers occurs at the rhizosphere zone within 1–3 mm area. Several interface reactions occur among the plant, soil biota, and other allied organisms in this region and govern the transformation of nanofertilizers/nano nutrients depending on the chemical situation in the rhizosphere area. Phytometallophores mainly consist of secondary metabolites and are classified into two groups on the basis of molecular weight. Mucilage creates a covering on nanofertilizers, which substantially modify their unique characteristics.

Low molecular weight organic compounds seem to influence chemical conversion of nano nutrients at rootzone. Exudation of organic acids creates acidic conditions surrounding the root zone as compared to the outside soil, thus increasing the dissolution rate of nanofertilizers. Similarly, oxidation reduction reactions may occur in the rhizosphere. Plant exudates consist of polyphenols and different acids which cause reduction of $CeO_2$ nanofertilizers and produce $Ce^{3+}$, which subsequently converts into $CePO_4$ or Ce carboxylates in the root zone and on the root (Zhang et al. 2017b). During interaction of Cu nanofertilizers with root exudates, a more intricated redox reaction can arise, which releases Cu with different valencies like (II), (I), and (0) in the rhizosphere (McManus et al. 2018). Root exudate composition varies with plant species. For instance, gramineous species specifically secrete mugineic acids, which aggravate the nanofertilizer dissolution process (McManus et al. 2018). The characteristics of root exudates secreted by cabbage (*Brassica oleracea var. gongylodes L.*) are different from those of the cucumber (*Cucumis sativus L.*) (Vančura and Hovadik 1965). Hence, the nature and degree of the alteration of nanofertilizers depends on the plant species. Moreover, at the leaf interface, different transformation processes including oxidation reduction reactions, aggregation, and dissolution may also occur. The interaction of nanofertilizers with dust or phyllospheric microorganisms causes their aggregation (Kim et al. 2010) and usually the sap released from the leaf or used in irrigation and rainfall change the state of aggregation. Plausibly, the epiphytes, colony of microorganisms including bacteria, fungi, and yeast, govern the chemical transformation of nanofertilizers at the phyllosphere (Vorholt 2012). Generally, microorganisms excrete extracellular polymeric substances, which alter the surface chemistry of nanofertilizers by triggering dissolution of the particle. A lack of clarity still exists on the influence of the phyllosphere in the conversion of nanofertilizers. Plant metabolites, like root exudates, play a key role during conversion of nanofertilizers at the interface of root-nano-nutrients. The existing knowledge base related to mineral-root exudate interactions may be extrapolated to understand the chemical behaviur of nanofertilizers with root exudates. Plants extracted an enormous amount of nutrients from the soil by releasing phytometallophores either by chelation or reduction in pH of rhizosphere. Typically, plants endure iron deficiency in soil by releasing different organic acids to increase the Fe-mineral dissolution (Jin et al. 2007). Experiment conducted under standard conditions recorded that some vital constituents, like various organic acids, (Manrique et al. 1991) sugars, and phenols, (Wang et al. 2012) facilitated the reduction and dissolution of nanofertilizers near the root-zone. Furthermore, the root exudate composition varies with plant species and is unique in nature. Hence, qualitative and quantitative estimation of the specific exudate components is essential for investigating nanofertilizer transformations by different plant species. The xylem and phloem fluids contain different metabolites, which are also significant in planta transformation of nanofertilizers, but the mechanism is still obscure. Transformation of nanofertilizers by various organic compounds and inorganic ions in the vascular fluids is facilitated. For illustration, due to the presence of organic constituents, two phenomena like dissolution and corona formation occur during nanofertilizer translocation within the xylem and phloem. Consequently, behavior and the fate of nanofertilizers may change during transportation inside the plant sap. Moreover, nanofertilizer alteration in plant leaf surface is also an important step, but is mostly unexplored. Noteworthy to mention that after entering into the plant cell and chloroplast, CNTs improve the photosynthesis rate (Giraldo et al. 2014). Nevertheless, functionality of CNTs may be lost over long duration interactions and also the safety of products changed from their original state may become a concern. Transformation of nanofertilizers is a dynamic and active process. Nanofertilizers may dissolve at the rhizosphere zone and enter into plant tissues where the prevailing reducing condition facilitates the restructuring of the nanoscale materials inside the tissue.

Rapid release of ions from ZnO and CuO nanofertilizers at the root surface and the subsequent instantaneous uptake of ions may cause a toxic effect on the plant. Fast transformations lead to the loss of the desired behavior and fate of nanofertilizers in soil matrix. Formation of corona on the surface of nanofertilizers modify material uptake momentarily but at a later stage release the nano nutrient inside the plant. The corona covers the particle surface partially and may be substituted by other molecules or liaised by proteins immediately.

Reaction rate for these transformations is crucial to comprehend the succeeding behavior and the ultimate state of nanofertilizers. This process leads to reshaping of nanofertilizers for precision agriculture as well as targeted or controlled release that represent these transformations at the strategy stage. The dynamic nature and heterogenicity of the soil matrix and ecology control the conversion of nanofertilizers. Nanofertilizers are categorized into various kinds based on bandgaps, size, shape, crystal structure, surface charge, etc., which can alter the conversion process. Soil chemical condition, the microbial community in soil, and the chemistry of root zone soil, as well as environment and climate conditions, together significantly govern the transformation of nanofertilizers. Furthermore, the contact of nanofertilizers with plant roots composition of phyto-metallophores like augmenting the excretion of different acids exudates and the chelation ability for elements. This process can sequentially again change alteration of nanofertilizers. Similarly, nanofertilizers also affect the microbial community at the rhizosphere zone, instigating variation in the microbial population and their activities, which subsequently alters nanofertilizer transformation. The soil microbiomes and plant activities are also interrelated. Additionally, parameters like climate change-induced elevated $CO_2$ content and variation in soil water content and heat capacity bring modifications in the microbial community structure and the secretion rate of exudates, and that in turn affect nanofertilizer transformations. Evidently, only a comprehensive method will help to comprehend the intricate transformation processes of nanofertilizers in the soil–plant system.

Experimental results revealed that zinc oxide nano particle application at the rate of 0.28 mg/L in a solution culture augmented the growth of maize plants in comparison to conventional zinc fertilizer as zinc sulphate (0.5 mg/L). The shoot and root dry matter weight (DMY) of a maize plant was enhanced by the application of Zn as $ZnSO_4$ and also nano ZnO (<100 nm). It is noteworthy to mention that shoot DMY 5.12 g/pot and root DMY 0.76 g/pot were observed at the dose of 0.5 mg Zn/L as ZnO nano particles (Adhikari et al. 2015). Hence it concluded that maize plants can uptake zinc, where zinc is applied through zinc oxide nano particles (<100 nm).

## 3.4 Epilogue

Once the manufactured nano-nutrients enter into the soil system, different dynamic physical and chemical environments change their size, shape, and concentration from their original state. Illustrating and envisaging environmental conversions can signify a test for numerous explanations. First, variation in the soil solution parameters, including pH, EC, oxidation and reduction state, and organic matter content may be responsible for diverse behaviors and consequently the types of transformations. Second, several transformations are dynamic and irreversible in nature which govern the interaction of the nano nutrient and the soil environment. Third, nano nutrients will probably undergo several alterations concurrently including reaction with organic matter and redox state, or consecutive non-reversible changes like irreversible sorption of various proteins on sequential contacts to intercellular materials and plasma. Thus, it is required to gather knowledge regarding the soil characteristics and the nature of the nano nutrients over a long period to predict how a specific nano nutrient will act in the soil system. Currently several advanced instruments are accessible to estimate and quantify the properties of nano nutrients, including surface charge, specific size, chemical composition, and concentration. Simultaneously specific protocols or methods should be developed to improve the tracking system dealing with the rate and extent of changes of nano nutrients under actual states, specifically in case of multiple transformations. A multidisciplinary approach like collaboration among the environmental scientists, environmental chemists, metrological and analytical researchers is urgently required for standard method development and validation. For instance, conversions of nano nutrients will increase over the next few months or years. To get insight into these transformations in totality will entail a skill to recuperate the nano nutrients from the soil and plant system and estimate the properties of these partially changed nano nutrients, or in situ condition through real time analysis. Though various studies have been conducted to investigate the fate and transport of manufactured nano enhanced materials, researchers may lay emphasis on clarifying the movement, translocation, rate of reaction, residence time in soil solution, and effects of interaction. Steady changes of nano enhanced materials will be essentially measured to comprehend the final fate and dispersal in the soil and biological system.

## References

Adamu G, Aliyu A (2012) Determination of the influence of texture and organic matter on soil water holding capacity in and around Tomas irrigation scheme, Dambatta local government Kano state. Res J Environ Earth Sci 4:1038–1044

Adhikari T, Kundu S, Biswas AK, Tarafdar JC, Subba Rao A (2015) Characterization of zinc oxide nano particles and their effect on growth of maize (Zea mays L.) plant. J Plant Nutr 38(10):1505–1515

Adhikari T, Sarkar D, Mashayekhi H, Xing B (2016) Growth and enzymatic activity of maize (Zea mays L.) plant: solution culture test for copper dioxide nano particles. J Plant Nutr 39(1):99–115
Allen BL, Kotchey GP, Chen Y, Yanamala NV, Klein-Seetharaman J, Kagan VE (2009) Mechanistic investigations of horseradish peroxidase-catalyzed degradation of single-walled carbon nanotubes. J Am Chem Soc 131(47):17194–17205
Aubert T, Burel A, Esnault MA, Cordier S, Grasset F, Cabello-Hurtado F (2012) Root uptake and phytotoxicity of nanosized molybdenum octahedral clusters. J Hazard Mater 219–220:111–118
Baalousha M (2017) Effect of nanomaterial and media physicochemical properties on nanomaterial aggregation kinetics. NanoImpact 6:55
Badawy AME, Luxton TP, Silva RG, Scheckel KG, Suidan MT, Tolaymat TM (2010) Impact of environmental conditions (pH, ionic strength, and electrolyte type) on the surface charge and aggregation of silver nanoparticles suspensions. Environ Sci Technol 44(4):1260
Banerjee P, Jain PK (2018) Mechanism of sulfidation of small zinc oxide nanoparticles. RSC Adv 8:34476
Bartlett RJ, James BR (1993) Redox chemistry of soils. Adv Agron 50:151–208
Brady NC, Weil RR (2008) The nature and properties of soils, vol 13. Prentice Hall, Upper Saddle River, NJ
Cervini-Silva J, Fowle DA, Banfield J (2005) Biogenic dissolution of soil cerium-phosphate minerals. Am J Sci 305:711–726
Chen C, Li YF, Qu Y, Chai Z, Zhao Y (2013) Advanced nuclear analytical and related techniques for the growing challenges in nanotoxicology. Chem Soc Rev 42:8266–8303
Conway JR, Adeleye AS, Gardea-Torresdey J, Keller AA (2015) Aggregation, dissolution, and transformation of copper nanoparticles in natural water. Environ Sci Technol 49:2749
Corbo C, Molinaro R, Parodi A, Toledano Furman NE, Salvatore F (2016) The impact of nanoparticle protein corona on cytotoxicity, immunotoxicity and target drug delivery. E. Tasciotti. Nanomedicine 11(1):81–100
Das P, Barua S, Sarkar S, Chatterjee SK, Mukherjee S, Goswami L, Das S, Bhattacharya S, Karak N, Bhattacharya SS (2018) Mechanism of toxicity and transformation of silver nanoparticles: inclusive assessment in earthworm-microbe-soil-plant system. Geoderma 314:73–84
David CA, Galceran J, Rey-Castro C, Puy J, Companys E, Salvador J, Monné J, Wallace R, Vakourov A (2012) Dissolution kinetics and solubility of ZnO nanoparticles followed by AGNES. J Phys Chem C 116:11758
Degenkolb L, Kaupenjohann M, Klitzke S (2019) The variable fate of ag and $TiO_2$ nanoparticles in natural soil solutions-sorption of organic matter and nanoparticle stability. Water Air Soil Pollut 230:62
Devi GP, Ahmed KBA, Varsha MKNS, Shrijha BS, Lal KKS, Anbazhagan V, Thiagarajan R (2015) Sulfidation of silver nanoparticle reduces its toxicity in zebrafish. Aquat Toxicol 158:149–156
Duan L, Hao R, Xu Z, He X, Adeleye AS, Li Y (2017) Removal of graphene oxide nanomaterials from aqueous media via coagulation: effects of water chemistry and natural organic matter. Chemosphere 168:1051–1057
Gardea-Torresdey J, Parsons J, Gomez E, Peralta-Videa J, Troiani H, Santiago P, Yacaman MJ (2002) Formation and growth of au nanoparticles inside live alfalfa plants. Nano Lett 2(4):397–401
Giraldo JP, Landry MP, Faltermeier SM, McNicholas TP, Iverson NM, Boghossian AA, Reuel NF, Hilmer AJ, Sen F, Brew JA (2014) Plant nanobionics approach to augment photosynthesis and biochemical sensing. Nat Mater 13:400–408
Gogos A, Voegelin A, Kaegi R (2018) Influence of organic compounds on the sulfidation of copper oxide nanoparticles. Environ Sci Nano 5:2560
Guo Z, Xie C, Zhang P, Zhang J, Wang G, He X, Ma Y, Zhao B, Zhang Z (2017) Toxicity and transformation of graphene oxide and reduced graphene oxide in bacteria biofilm. Sci Total Environ 580:1300–1308

Jahn R, Blume HP, Asio VB, Spaargaren O, Schad P (2006) Guidelines for soil description, 4th edn. FAO, Rome
Jin CW, You GY, He YF, Tang C, Wu P, Zheng SJ (2007) Iron deficiency-induced secretion of phenolics facilitates the reutilization of root apoplastic iron in red clover. Plant Physiol 144:278–285
Kah M, Tufenkji N, White JC (2019) Nano-enabled strategies to enhance crop nutrition and protection. Nat Nanotechnol 14(6):532–540
Kampe S, Kaegi R, Schlich K, Wasmuth C, Hollert H, Schlechtriem C (2018) Silver nanoparticles in sewage sludge: bioavailability of sulfidized silver to the terrestrial isopod Porcellio scaber. Environ Toxicol Chem 37:606
Kim H-S, Ahn J-Y, Hwang K-Y, Kim I-K, Hwang I (2010) Atmospherically stable nanoscale zero-valent iron particles formed under controlled air contact: characteristics and reactivity. Environ Sci Technol 44:1760–1766
Koelmel J, Leland T, Wang H, Amarasiriwardena D, Xing B (2013) Investigation of gold nanoparticles uptake and their tissue level distribution in rice plants by laser ablation-inductively coupled-mass spectrometry. Environ Pollut 174:222
Kumar N, Pacheco JL, Noël V, Dublet G, Brown GE (2018) Sulfidation mechanisms of Fe(iii)-(oxyhydr)oxide nanoparticles: a spectroscopic study. Environ Sci Nano 5:1012
Lewis RW, Bertsch PM, McNear DH (2019) Nanotoxicity of engineered nanomaterials (ENMs) to environmentally relevant beneficial soil bacteria - a critical review. Nanotoxicology 13(3):392–428
Li M, Wang P, Dang F, Zhou D-M (2017) The transformation and fate of silver nanoparticles in paddy soil: effects of soil organic matter and redox conditions. Environ Sci Nano 4:919–928
Lin D, Story SD, Walker SL, Huang L, Q , W. & Cai, P. (2017) Role of pH and ionic strength in the aggregation of TiO2 nanoparticles in the presence of extracellular polymeric substances from Bacillus subtilis. Environ Pollut 228:35–42
Lowry GV, Gregory KB, Apte SC, Lead JR (2012) Transformations of nanomaterials in the environment. Environ Sci Technol 46:6893–6899
Luo M, Huang Y, Zhu M, Tang Y-N, Ren T, Ren J, Wang H, Li F (2018) Properties of different natural organic matter influence the adsorption and aggregation behavior of $TiO_2$ nanoparticles. J Saudi Chem Soc 22:146–154
Lynch I, Weiss C, Valsami-Jones E (2014) A strategy for grouping of nanomaterials based on key physico-chemical descriptors as a basis for safer-by-design NMs. Nano Today 9:266–270
Ma R, Levard C, Michel FM, Brown GE Jr, Lowry GV (2013) Sulfidation mechanism for zinc oxide nanoparticles and the effect of sulfidation on their solubility. Environ Sci Technol 47:2527
Manrique LA, Jones CA, Dyke PT (1991) Predicting cation-exchange capacity from soil physical and chemical properties. Soil Sci Soc Am J 55:787–794
McManus P, Hortin J, Anderson AJ, Jacobson AR, Britt DW, Stewartm J, McLean JE (2018) Rhizosphere interactions between copper oxide nanoparticles and wheat root exudates in a sand matrix: influences on copper bioavailability and uptake. Environ Toxicol Chem 37:2619–2632
Ohshima H (2012) Electrical phenomena at interfaces and biointerfaces: fundamentals and applications in nano-, bio-, and environmental sciences. John Wiley & Sons, New York, NY, p 27
Pearson RG (1963) Hard and soft acids and bases. J Am Chem Soc 85:3533
Ren M, Zhang Z, Wang X, Zhou Z, Chen D, Zeng H, Zhao S, Chen L, Hu Y, Zhang C (2018) Diversity and contributions to nitrogen cycling and carbon fixation of soil salinity shaped microbial communities in Tarim Basin. Front Microbiol 9:431
Roy P, Das B, Mohanty A, Mohapatra S (2017) Green synthesis of silver nanoparticles using Azadirachta indica leaf extract and its antimicrobial study. Appl Nano sci 7:843–850
Rui Y, Zhang P, Zhang Y, Ma Y, He X, Gui X, Li Y, Zhang J, Zheng L, Chu S, Guo Z, Chai Z, Zhao Y, Zhang Z (2015) Transformation of ceria nanoparticles in cucumber plants is influenced by phosphate. Environ Pollut 198:8–14
Salas EC, Sun Z, Luttge A, Tour JM (2010) Reduction of graphene oxide via bacterial respiration. ACS Nano 4:4852

Schauer N, Fernie AR (2006) Plant metabolomics: towards biological function and mechanism. Trends Plant Sci 11(10):508–516

Schnitzer M, Khan SU (1975) Soil organic matter, vol 8, 1st edn. Elsevier, Amsterdam

Sharma JK, Akhtar MS, Ameen S, Srivastava P, Singh G (2015) Green synthesis of CuO nanoparticles with leaf extract of Calotropis gigantea and its dye-sensitized solar cells applications. J Alloys Compd 632:321–325

Shetty K (2004) Role of proline-linked pentose phosphate pathway in biosynthesis of plant phenolics for functional food and environmental applications: a review. Proc Biochem 39:789–803

Singh A, Hou W-C, Lin T-F, Zepp RG (2019) Roles of silver–chloride complexations in sunlight-driven formation of silver nanoparticles. Environ Sci Technol 53(19):11162–11169

Solanki P, Bhargava A, Chhipa H, Jain N, Panwar J (2016) Nano-fertilizers and their smart delivery system. In: Rai M, Ribeiro C, Mattoso L, Duran N (eds) Nanoscience in food and agriculture. Springer International Publishing AG, Cham, pp 81–101. ISBN 978-3-319-14023-0

Tang Z, Cheng T (2018) Stability and aggregation of nanoscale titanium dioxide particle ($nTiO_2$): effect of cation valence, humic acid, and clay colloids. Chemosphere 192:51–58

Tang F, Uchikoshi T, Sakka Y (2002) Electrophoretic deposition behavior of aqueous nanosized zinc oxide suspensions. J Am Ceram Soc 85:2161

Torney F, Trewyn BG, Lin VSY, Wang K, (2007) Mesoporous silica nanoparticles deliver DNA and chemicals into plants. Nature Nanotech 2:295–300

Valsami-Jones E, Lynch I (2015) Nanosafety. How safe are nanomaterials. Science:350–388

Vančura V, Hovadik A (1965) Root exudates of plants. Plant Soil 22:21–32

Vorholt J (2012) Microbial life in the phyllosphere. Nat Rev Microbiol 10:828–840

Wang Z, Xie X, Zhao J, Liu X, Feng W, White JC, Xing B (2012) Xylem- and phloem-based transport of CuO nanoparticles in maize (*Zea mays* L.). Environ Sci Technol 46(8):4434–4441

Wang F, Wang F, Gao G, Chen W (2015) Transformation of graphene oxide by ferrous iron: environmental implications. Environ Toxicol Chem 34:1975

Wang W, Yang Z, Zhang A, Yang S (2021) Water retention and fertilizer slow release integrated superabsorbent synthesized from millet straw and applied in agriculture. Ind Crop Prod 160:113126

Xiao H, Liu N, Tian K, Liu S, Ge F (2018) Accelerated effects of nano-ZnO on phosphorus removal by Chlorella vulgaris: formation of zinc phosphate crystallites. Sci Total Environ 635:559–566

Xie C, Zhang J, Ma Y, Ding Y, Zhang P, Zheng L, Chai Z, Zhao Y, Zhang Z, He X (2019) Bacillus subtilis causes dissolution of ceria nanoparticles at the nano–bio interface. Environ Sci Nano 6:216–223

Zhang P, Ma Y, Zhang Z, He X, Zhang J, Guo Z, Tai R, Zhao Y, Chai Z (2012a) Biotransformation of ceria nanoparticles in cucumber plants. ACS Nano 6(11):9943–9950

Zhang P, He X, Ma Y, Lu K, Zhao Y, Zhang Z (2012b) Distribution and bioavailability of ceria nanoparticles in an aquatic ecosystem model. Chemosphere 89:530–535

Zhang P, Xie C, Ma Y, He X, Zhang Z, Ding Y, Zheng L, Zhang J (2017a) Shape-dependent transformation and translocation of ceria nanoparticles in cucumber plants. Environ Sci Technol Lett 4:380–385

Zhang P, Ma Y, Liu S, Wang G, Zhang J, He X, Zhang J, Rui Y, Zhang Z (2017b) Phytotoxicity, uptake and transformation of nano-CeO2 in sand cultured romaine lettuce. Environ Pollut 220:400–1408

Zhang P, Misra S, Guo Z, Rehkämper M, Valsami-Jones E (2019) Stable isotope labeling of metal/metal oxide nanomaterials for environmental and biological tracing. Nat Protoc 14(10):2878–2899

Zhou JM, Huang PM (2007) Kinetics of potassium release from illite as influenced by different phosphates. Geoderma 138:221–228

# Chapter 4
# Nanofertilizers Use in India and in the World

**Rahul Mishra, Dhiraj Kumar, Nishant Kumar Sinha, S. P. Datta, Jitendra Kumar, Binoy Sarkar, and Ashok K. Patra**

**Abstract** The excess use of chemical fertilizers creates residues in the agricultural produce, pollutes the environment, upsets the ecosystems, and deteriorates soil quality and health. Moreover, an excess amount of nutrients gets lost from the soil due to poor use efficiency and faulty methods of nutrient delivery. Controlled release of nutrients through a nanotechnological approach thus becomes cost-effective and efficient over conventional fertilizers. Nanofertilizers consist of one or more of the plant nutrients wherein at least more than 50% of the particles are smaller than 100 nm in size. Appropriate formulation of nanofertilizers tends to release the nutrients slowly in a steady state so that nutrients are released over an extended period. Apart from reducing nutrient loss and increasing nutrient use efficiency, nanofertilizers activate plant defense mechanisms against insects, pests, and other associated environmental stresses. Nanofertilizers hold the potential to improve environmental sustainability, but current research on the topic is mainly limited to controlled environmental conditions. Large-scale production, material availability, and appropriate field application logistics, especially in developing countries, are still lagging. Future research is needed to understand the long-term impact of nanofertilizers under field conditions, especially for major nutrients, in crop growth, soil quality, and overall environmental sustainability.

**Keywords** Nanofertilizers · Crop cultivation · Controlled nutrient release · Nutrient use efficiency · Soil and water quality

---

R. Mishra (✉) · D. Kumar · N. K. Sinha · S. P. Datta · J. Kumar · A. K. Patra
ICAR-Indian Institute of Soil Science, Bhopal, Madhya Pradesh, India

B. Sarkar
Future Industries Institute of University of South Australia, Adelaide, Australia
e-mail: Binoy.Sarkar@unisa.edu.au



A. K. Patra (ed.), *Nanofertilizers in Agriculture*, Sustainability Sciences in Asia and Africa, https://doi.org/10.1007/978-981-97-8112-6_4

## 4.1 Introduction

To address the global issues pertaining to environmental crises, shortage of food and nutritional insecurities, there is an urgent need for action through both scientific research as well as policy interventions (Adhikari and Ramana 2019; Aguilar-Pérez et al. 2020; Bilal et al. 2021). The increase in environmental pollutants associated with the expansion of the population and industrial advancements (Qamar et al. 2022). It is predicted that the global population may attain the 11 billion mark by 2100 (Adam 2021). These calls for an immediate need to enhance current food production by 60–70% in order to meet rising calorie requirements of the expanding populace (Van Dijk et al. 2021). Thus, it is essential to strategically increase resource utilization efficiency while sustaining the agricultural production in a robust way through the application of modern science and technology. In developing countries, around 50–55% of the increment in crop yield is often associated with the application of chemical fertilizers (Adhikari and Ramana 2019). However, the nutrient utilization efficiency from these fertilizers remains quite low, with nitrogen (N) utilization ranging from 30% to 40%, phosphorus (P) from 15% to 20%, potassium (K) from 50% to 55%, and micronutrients at 2%–5% (Adhikari and Ramana 2019). The extensive application of imbalance dose of chemical fertilizers resulted in excessive mining of soil nutrients, thus causing a net negative balance of soil nutrients of around ten million tonnes and thus contributes towards deterioration of soil health (Chugh et al. 2021). Although the use of chemical fertilizers is instrumental in supporting food production for the global population, their sustainability is now being questioned (Molina-Santiago and Matilla 2020). These fertilizers not only cause environmental harm (Rosa and Gabrielli 2022; Kopittke et al. 2019) but also play a role in greenhouse gas (GHGs) emissions (Kyriakou et al. 2020). Further, the runoff from agricultural areas, containing chemical fertilizers, poses a threat to surface and groundwater, thus impacting native ecosystems (Vitousek et al. 1997; Dirzo and Raven 2003; Owuor et al. 2011; Cheng et al. 2015). Apart from the environmental concerns, the usage of chemical fertilizers imposes economic burdens, constituting between 20% and 36% of a farm's operating budget (Schnitkey et al. 2021). During the last 2 years, the cost of chemical fertilizers nearly tripled, surging from \$94 per acre to \$232 per acre of corn (Komarek et al. 2017; Brownlie et al. 2023). This hike in costs not only impacts consumer food prices but has also the potential to decrease yields due to a sub-optimal dose of chemical fertilizer application (Schnitkey et al. 2021).

The environmental and economic challenges associated with traditional agricultural practices emphasize the necessity for alternative solutions. With agricultural land being a finite resource, ensuring its long-term sustainability is of paramount importance for global food security. In order to make intensive agriculture productive as well as environment friendly, systematic initiatives should be implemented for restoring the natural resources. Achieving sustainability in agriculture involves adopting innovative techniques (Shang et al. 2019) that not only lead to farm production increase but also mitigate the adverse environmental impacts (Arora 2018).

Concerning sustainable agriculture, the area of nanotechnological interventions, especially for fertilizer development, emerges as a promising approach for significant increase in crop production as well as meeting the demands of an ever-growing world population (Lal 2008). Acknowledging very limited efficiency (only 30–50%) of traditional fertilizers and the paucity of management options for enhancing their rates, DeRosa et al. (2010) advocated the integration of nanotechnology into the area of fertilizer research and development. While numerous experts from the agricultural sector have expressed interest in the development and application of nanomaterial-related fertilizers (Lal 2008; Nair et al. 2010; Ghormade et al. 2011; Khot et al. 2012), direct research in this field is still lagging or is in its early stages. However, recent advancements in the area of nanotechnology have clearly showcased the promising potential for development of nanofertilizers and their application. Currently, the use of nano hybrid constructs, such as nanofertilizers (NFs), has gained significant attention for achieving sustainable crop yields along with preserving the quality of the environment via smart agrochemical delivery (Raliya et al. 2017a, b; Okey-Onyesolu et al. 2021).

### *4.1.1 Definition or Concept of Nanofertilizer/Nanomaterials*

Nanomaterials are materials characterized by a single unit size ranging between 1 and 100 nanometers (nm) in at least one dimension. In the context of agriculture, nanofertilizers encompass nanomaterials that function either as supplier of nutrients, thereby enhancing plant growth and yields, or as boosters for traditional fertilizers, without directly supplying nutrients to the crops. These nanofertilizers are often classified into macronutrient, micronutrient, polymer-based, and trace element-based. In comparison to traditional fertilizers, there is the expectation of notable improvements in crop growth and productivity through the application of NFs. They carry enough potential to optimize use efficiency of fertilizers, diminish the nutrient losses, and mitigate adverse environmental impacts (Liu and Lal 2015). This implies that the utilization of nanofertilizers holds the prospect of addressing challenges linked to conventional fertilization methods in agriculture.

### *4.1.2 Synthesis of Nanofertilizers*

The predominant technique for developing nanofertilizers (NFs) through nanomaterials is nutrient encapsulation, marked by recent progress, where specific nutrients are enclosed within nano porous materials, thin polymer film particles as coating material, or included in nanoscale emulsions. The encapsulation of beneficial microbes, such as bacteria or fungi, has shown promise in enhancing the N, P, and K availability around the root zone (Bose 2020). NFs can be synthesized from standard fertilizer bulk ingredients after undergoing the process of extraction and

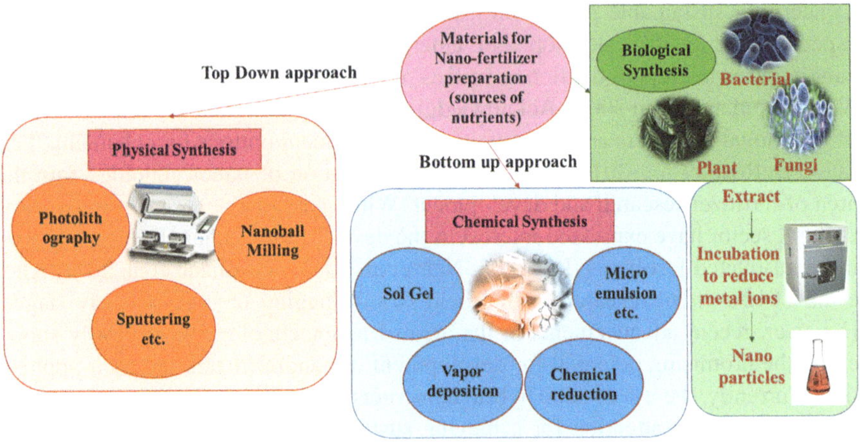


**Fig. 4.1** Process of synthesis of nanofertilizers (Babu et al. 2022)

purification, employing several chemical, physical, mechanical, and biological methods (Fig. 4.1). The initial creation of nanomaterials involves mainly two approaches: top-down and bottom-up. The top-down method includes techniques like ball milling, photolithography, sputtering, chemical etching, and also other physical methods, while the bottom-up approach utilizes chemical and biological methods, involving atomic and molecular compression, spray, laser pyrolysis (photochemical, sonochemical, chemical and electrochemical, microwave, sol-gel, vapor deposition, etc.). Among these, physical methods such as ultra-sonication, mechanical pressure, or electrical energy are considered as low-toxic, cost-effective, efficient, and eco-friendly for producing nanoparticles (NPs) (Satyanarayana and Reddy 2018). Nowadays biological methods are emerging as a sustainable method for synthesizing NFs, offering benefits such as time and energy efficiency, economic viability, and the use of sustainable, environmentally clean, and non-hazardous components. In laboratories, NFs are synthesized using several biological and precursor materials, including fungi, bacteria, and plants. For example, during the production of selenium nanoparticles (Se-NPs), compounds like selenous acid, sodium selenite, or selenium oxide can act as precursors, while plant substrates like bougainvillea wildflowers, grapefruits, and others were employed (Sharma et al. 2014; Ganesan 2015).

### *4.1.3 Nanofertilizer Use in Crop Production*

Various forms of nutrient carriers exist in NFs, categorized according to their nutrient composition. These include categories based on their macronutrient, micronutrient, carbon, and polymer carriers (Fig. 4.2). The adoption of using nanofertilizers (Jha et al. 2023) may serve as a big boost toward fulfilling the very objective of sustainable agriculture (Mahapatra et al. 2022) in India, as well as across the world,

**Fig. 4.2** General classes of nanofertilizers. (Modified from Liu and Lal 2015)

by means of curtailing the doses of fertilizers (Kumar et al. 2021; Upadhyay et al. 2023a), reduction of runoff and leaching, as well as atmospheric gaseous emissions (Manjunatha et al. 2016). Several studies have shown that nano urea sprayed crops led to improvement in the crop yield under field conditions (Das et al. 2016; Manikandan and Subramanian 2016; Raliya et al. 2017a, b; Du et al. 2019; Rathore et al. 2019; Kumar et al. 2021; Upadhyay et al. 2023a). Impact of nano- macro- and micronutrient fertilizers on the crop growth and yield parameters were summarized in Table 4.1. Macronutrient NFs consist of one or more chemical elements that are essential for plants, like N, P, K, Mg and Ca. These nanofertilizers are capable of providing these vital macronutrients to the plants. The worldwide usage of macronutrient fertilizers ($N + P_2O_5 + K_2O$) was to the tune of 175.7 Mt. in 2011 and is expected to rise to 263 Mt. by 2050 (Alexandratos and Bruinsma 2012). The integration of nano-N and nano-Zn with traditional N fertilizers has tremendous potential to improve the crop yields, nutrient uptake, soil mineral-N, soil microbial

**Table 4.1** Different nanofertilizers used in crop production and their response

| Element supplied | Nanofertilizer | Response | Reference |
|---|---|---|---|
| Phosphorus (P) | Hydroxyapatite ($Ca_5(PO_4)_3OH$); dose of 21.8 mg $L^{-1}$ as P | Soybean; enhancement in growth rate @ 33% and seed yield @ 20%, respectively | Liu and Lal (2014) |
| Potassium (K) | K nano particle synthesized from potassium nitrate; dose of 150 mg $L^{-1}$ as K | Groundnut; increase nutrient content in shoots and seeds of plants | Afify et al. 2019 |
| Calcium (Ca) | Calcium carbonate ($CaCO_3$); dose of 160 mg $L^{-1}$ as Ca | Groundnut; improve plant growth and biomass yield | Liu et al. (2005) |
| Magnesium (Mg) | Mg-NP; dose of 0.5 g $L^{-1}$ of Mg-NP | Black-eyed pea (*Vigna unguiculata*); enhance seed weight | Delfani et al. (2014) |
| Iron (Fe) | Superparamagnetic Fe-NPs; 30, 45 and 60 mg $L^{-1}$ | Soybean; reduce chlorotic symptom | Ghafariyan et al. (2013) |
| | Fe-NPs; dose of 500 mg $L^{-1}$ as Fe | Black-eyed peas; enhance number of pods | Delfani et al. (2014) |
| Zinc (Zn) | ZnO NP; dose of 1–2000 mg $L^{-1}$, | Mung bean and chickpea; enhance shoot and biomass yield | Mahajan et al. (2011) |
| Copper (Cu) | Cu- NPs; dose of <0.25 mg $L^{-1}$ as Cu | Waterweed (Elodea. Densa planch); enhance photosynthesis rate | Nekrasova et al. (2011) |
| | Metallic Cu-NPs; dose of 130 and 600 mg $kg^{-1}$ | Lettuce; significant enhancement in seedling growth | Shah and Belozerova (2009) |
| | Metallic Cu-NPs; dose of 200–1000 mg $kg^{-1}$ | Significant reduction in mung bean, wheat and yellow Squash growth | Lee et al. (2008), Musante and White (2012) |
| Molybdenum (Mo) | Mo-NPs; dose of 8 mg $L^{-1}$, seed-soaked 1–2 h | Chickpea; enhance nodulation and antioxidant enzyme, microbial activity | Taran et al. (2014) |

biomass carbon, and dehydrogenase activity in wheat-maize and mustard-pearl millet cropping sequence, maximum grain yield of wheat, maize, and pearl millet and mustard crops found under $RRN_{100}PK$ + Nano-N + Nano-Zn treatments. Basal application of N (75% of recommended) through the use of prilled urea with a full dose of $P_2O_5$ and $K_2O$, along with nano-urea (2500 mL $ha^{-1}$) + nano-Zn (1250 mL $ha^{-1}$) sprays, resulted in at par grain yield (wheat, maize, mustard, and pearl millet) compared to 100% N+ full dose of $P_2O_5$ and $K_2O$ recommended dose of fertilizers (Upadhyay et al. 2023b). Further, it was found that the N, P, K, and Cu(%) and Zn (ppm) content in grain and straw were maximum in grain and straw under 50% N and 0% Zn, 100% P and K + 2 sprays of nano-N mixed with nano-Zn and Nano-Cu in central Uttar Pradesh (Yadav et al. 2021a, b). In another study done at sub-humid southern plains of Rajasthan, it was observed that Nano P, K and Zn, when applied through foliar spray, manipulate the growth parameters of wheat, which results in a significant improvement in yield and yield related parameters. The increased growth may be attributed to enhanced growth hormone and photosynthetic activities owing to applications of NFs. It was also observed that the application of N, P, and K in recommended doses, along with nano urea @ 4 ml, resulted in achieving the best physico-chemical properties of soil, which led to the maize yield increasing to 5330.04 kg $ha^{-1}$ (Rathore et al. 2022). One of the field experiments conducted at Garacharma, ICAR-CIARI, Port Blair, to investigate the impact of nano urea (2% and 4%) on the growth and productivity of okra demonstrated that utilization of nano spray resulted in a significant increase in okra yield (9.6%) when compared to application of NPK using chemical fertilizers (Subramani et al. 2023). The findings from 187 experiments done on farmers' potato fields have shown that the amount of urea used by farmers in order to provide N to the crop may be effectively reduced to half. The crop yields achieved by using 50% less N compared to the Farmer's Fertiliser Practice (FFP) and with the application of 2 sprays of nano-N in the growing crops were greater compared to yields with FFP in the majority of the experiments conducted (Tiwari et al. 2021). In 160 trials of organic farming, using nano N and Zn on wheat, pearl millet, and oil seed crops like mustard and sesame revealed that the mean yields of wheat, pearl millet, mustard, and sesame were enhanced by factors of 5.35, 4.22, 8.40, and 24.2%, respectively (Kumar et al. 2022).

Nano DAP meets chlorophyll's phosphorus needs for photosynthetic action. Nano DAP comprises of nano-sized (<100 nm) polymer-encapsulated DAP particles. This nanoparticle suspension in water may be sprayed on plant leaves after dilution. Nano DAP's tiny size (<100 nm) and large surface area enable plant leaf absorption. This new nano-formulation improves agricultural productivity, environmental impact, and farmer profitability (https://www.coromandel.biz/press-release/nano-dap/#:~:text=The%20small%20size%20of%20Nano,burden%20and%20increased%20farmer%20profitability). In kharif 2021, a field experiment at Jagdalpur, Chhattisgarh, examined the effects of Nano-DAP on nutrient use efficiency of rice and its yield. Results showed that 50% P + 100% NK + root dipping as well as two foliar sprays of Nano-DAP at 20–25 DAT and 45–50 DAT improved nutrient use efficiency and yield (Deo et al. 2022). Singh et al. (2021) found that nano-diammonium phosphate (n-DAP) outperformed conventional granular DAP (c-DAP) in wheat and tomato growth. However, additional research and field trials are necessary to completely decipher the knowledge on potential benefits and the long-term impact of nanofertilizers, both individually and in combination with conventional fertilizers. This may open up the way for the development of environmentally sustainable and more efficient crops. An example of macronutrient nanofertilizer is hydroxyapatite ($Ca_5(PO_4)_3OH$) nanoparticle (16 nm), which was used to evaluate its impact on the growth performance of soybean plants (*Glycine max*) in an inert growing medium consisting of 50% perlite and 50% peat moss under greenhouse experiment. The data indicated an increment of 33% in growth rate and 20% in seed production on utilization of NPs compared to the use of a conventional P fertilizer (Liu and Lal 2014). A further study revealed that the apatite nanoparticles have significantly lower affinities towards soil constituents compared to the ions i.e.,

($PO_4^{3-}$, $HPO_4^{2-}$, $H_2PO^{4-}$, or $Ca^{2+}$). Hence, a considerable proportion of NPs persist in the soil solution, allowing the roots to absorb them, whereas the majority of phosphate ions coming from traditional P fertilizers tend to be absorbed by soil particles and are not readily accessible to be up taken by plants. Furthermore, the apatite nanoparticles may have lower bioavailability to the harmful alga compared to soluble phosphate ions, resulting in less eutrophication hazards (Liu and Lal 2015). Potassium (K), one of the major important nutrients, imparts significant effect on crop production and also increases resistance to biotic as well as abiotic stress. The utilization of K-NFs @ 150 mg $kg^{-1}$resulted in enhanced nutrient uptake by the peanut plant (Afify et al. 2019). In a study done by Liu et al. (2005), using calcium carbonate nanoparticles (20–80 nm, 160 mg $L^{-1}$ as Ca) on the peanut seedlings that were grown in sand with Hoagland solution for about 80 days, findings showed that the Ca-NPs application significantly enhanced the growth and biomass yield of the seedlings (4.42 g per plant to 5.07 g per plant, 15% increment) compared to the control (no calcium application). However, improvements in dry biomass were more or less similar to those observed with the application of soluble calcium (200 mg $L^{-1}$ $Ca(NO_3)_2$ as Ca). In a study, Delfani et al. (2014) examined the magnesium nanoparticle (Mg-NP) and iron nanoparticle (Fe-NP) application impact on black-eyed pea plants (*Vigna unguiculata*) through foliar application. They found that a combined application of 0.5 g $L^{-1}$ of Mg-NP and Fe-NP increased the 1000 seeds weight by 7%, and this increase was also greater than the weight of seeds when regular Mg and Fe were applied.

Micronutrients are the elements that are required in smaller quantities compared to macronutrients but are essential for plant growth as mentioned in Hoagland solution composition (Hoagland and Arnon 1950). The most common micronutrient fertilizers used in agriculture are zinc (Zn), copper (Cu), iron (Fe), manganese (Mn), and molybdenum (Mo). Micronutrients are commonly incorporated into composite fertilizers, which consist of N, P, and K, at lower concentrations (<5.0 mg $L^{-1}$) as soluble salts. This helps in efficient absorption by crops. In their study, Ghafariyan et al. (2013) found that even small amounts of superparamagnetic Fe-NPs had a significant positive effect on the chlorophyll content in the sub-apical leaves of soybeans. This greenhouse experiment, conducted using hydroponic conditions, suggests that soybeans can effectively utilize Fe-NPs as a source of iron, thereby alleviating the iron deficiency-induced chlorosis symptoms. The utilization of Fe-NPs had a comparable effect to that of a potent Fe source for the plants, specifically Fe-EDTA (45 mg $L^{-1}$ as Fe content). Further, Delfani et al. (2014) found that applying 500 mg $L^{-1}$ of Fe-NPs to black-eyed peas through foliar application resulted in considerable improvements in the number of pods (47%), weight of 1000-seeds (7%), Fe content in leaves (34%), and chlorophyll content (10%). The utilization of Fe-NPs demonstrated a greater enhancement in crop performance compared to the utilization of conventional Fe salt. For Zn, the most common used nanomaterial is zinc oxide (ZnO) and their effects on plants are currently being extensively investigated. Mahajan et al. (2011) found that the development of mung bean and chickpea (*Cicer arietinum*) seedlings enhanced by application of

ZnO-NPs at low concentrations. The most favorable growth was observed in mung bean seedling development in terms of root length (42% enhanced) and root biomass (41% enhanced), as well as shoot length (98% enhanced) and shoot biomass (76% enhanced), when exposed to a concentration of 20 mg $L^{-1}$ over control. Copper nanoparticles (Cu-NPs) at concentrations below 0.25 mg $L^{-1}$ as Cu increased the plant photosynthetic rate by 35% over control (without Cu) (Nekrasova et al. 2011). Shah and Belozerova (2009) have shown that the application of metallic Cu-NPs to the soil at concentrations of 130 and 600 mg $kg^{-1}$ has resulted in a considerable improvement in lettuce seedling development after 15 days by 40% and 91%, respectively. However, elevated concentrations (ranging from 200 to 1000 mg $L^{-1}$) of metallic Cu-NPs were found to have detrimental impacts on the growth of seedlings of wheat, mung bean, and yellow squash (Lee et al. 2008; Musante and White 2012). Taran et al. (2014) use solution of Mo-NPs for seed treatment of chickpea for 1–2 h with treatment combination of (water, Mo-NPs, microbial incubation with N-fixing bacteria, and a combination of the microbes and Mo-NPs) and observed that Mo-NP treatment demonstrated direct benefits in the development of microorganisms and growth of chickpeas. Tarafdar et al. (2012) reported that biosynthesized Mo-NPs (*Aspergillus tubingensis* TFR 29; 2–7 nm at concentration of 4 ppm) resulted in a considerable enhancement of beneficial enzymes, root growth, and microbial activities in the rhizosphere. The intake of Ni is crucial for several enzymatic activities, the preservation of redoc conditions of the cell, and various other developmental processes, including biochemical, physiological, and growth responses (Yusuf et al. 2011). The addition of Ni-NFs (5 nm) @ 0.01 mg $L^{-1}$ boosted the concentration of chlorophyll 'a' and chlorophyll 'b' in wheat crops (Zotikova et al. 2018).

Nanomaterial enhanced fertilizers are used for crop production, generally defined as nanomaterials combined with plant nutrients, and enhance the nutrient uptake efficiency of plants and minimize the negative environmental impact. However, these nanomaterials do not contain or supply the specific nutrients themselves. Nutrient-augmented zeolites are the most common sort of this category of fertilizer. Perrin et al. (1998) observed the effect of $NH_4^+$ zeolite on maize growth and nitrogen use efficiency and reported that nitrogen use efficiency was enhanced; however, no significant effect on growth parameter was observed among the nitrogenous fertilizer. According to Deshpande et al. (2017), when Zn-chitosan nanoparticles are given to wheat plants in a foliar manner, they may enhance plant development. Similarly, Cu-Chitosan NP application enhances the development of maize and tomato plants (Saharan et al. 2016).

Multiple researchers showed that application of some of the nanoparticles (NPs) may also improve plant development to some extent, even though these particles do not possess any macro or micronutrients (Liu and Lal 2015). These include carbon nanotubes (CNTs) and $TiO_2$ nanoparticles (Ti-NPs). Conventionally, Titanium (Ti) is not considered to be a fundamental plant nutrient. In addition, majority of the soils have large amounts of Ti, between 0.1% and 0.9%, and the soil solution has about 0.03 mg $L^{-1}$ Ti by volume (Kabata-Pendias and Pendias 1984). Effect of Ti nanoparticle on crop agronomic parameters was evaluated by several researchers.

Yang et al. (2007) treated spinach (*Spinacia oleracea*) seeds in a solution containing 2.5 g $L^{-1}$ of titanium nanoparticles for a period of 2 days in the light. After this treatment, the seeds were planted in a growth medium and a solution was applied to the sprouted seedlings once in a week in a greenhouse for a period of 35 days through the foliar method. The results indicated that fresh and dry weights were almost twice as large compared to control (no $TiO_2$ NPs applied). The levels of total N, chlorophyll, as well as protein in leaves, exhibited a respective rise of 23%, 34%, and 13% over control. Song et al. (2012) observed that the addition of 0.5 g $L^{-1}$ of Ti-NPs led to a significant rise in the growth parameters of duckweed (*Lemnaminor*); however, concentrations of over 0.5 g $L^{-1}$ have shown a suppressive impact on this plant. Several studies have shown that carbon nanotubes (CNTs) at low concentrations may enhance the process of seed germination and promote plant development. In their study, Lin and Xing (2008) found that applying 2 g $L^{-1}$ of multi-walled carbon nanotubes (CNTs) resulted in a substantial enhancement in root elongation of germinated ryegrass seeds. However, the treatment of CNTs did not have a statistically significant impact on the root elongation of germinated rape and maize seeds compared to the control.

## 4.2 Commercial Scale Available Nanofertilizers

Commercially available NFs are gaining popularity, with numerous products now on the market (Table 4.2). Various companies produce these products, offering a range of formulations. Popular NFs include zinc oxide, silica, and iron oxide-based varieties, each providing specific nutrients to enhance crop production. Among these, NanoGro™, developed by Bayer CropScience (Pettersen and Aamlid 2019), is notable for reducing the need for fertilizers by increasing their efficiency and enabling more targeted application. NanoGro™ contains nanoscale particles that penetrate the soil to reach plant roots, promoting more effective fertilizer uptake. It comes in various formulations tailored to different crops and soil conditions. IFFCO Nano-urea from India is used as a source of nitrogen that enhances crop yield when applied in combination with conventional fertilizer (Upadhyay et al. 2023b). Another nanofertilizer is NanoFert, produced by Geolife® from India (Oancea 2022). This product enhances the effectiveness of traditional fertilizers by improving nutrient absorption. Like NanoGro™, NanoFert's nanoscale particles penetrate the soil to facilitate efficient nutrient uptake. NanoFert is available in multiple formulations for different crops and soil types. Geolife® also offers NanoFert Plus, which further boosts traditional fertilizer efficacy by aiding nutrient absorption through its nanomaterials. This product, too, is designed to penetrate the soil surface and reach plant roots, and it comes in various formulations suitable for different crops and soil conditions. IFFCO Nano-di ammonium phosphate (DAP) from India

**Table 4.2** Different nanofertilizers available in market

| S. no. | Nanofertilizer trade name | Composition and manufacturer | References |
|---|---|---|---|
| 1.0 | Nanogro™ | NPK with bio-stimulant; aqua-yield operations, LLC | Pettersen and Aamlid (2019) |
| 2.0 | NanoPro™ | Humic acid with silicon dioxide; aqua-yield operations, LLC | Espevig et al. (2020) |
| 3.0 | Magic green | Oxides of Ca and Mg with $SiO_2$ | Ekinci et al. (2014) |
| 4.0 | NovaLand Nano | Nitrogen, micronutrients with nanocomposites; Land Green & Technology Co. (Taiwan) | Menossi (2022) |
| 5.0 | IFFCO Nano-urea | 4.0% nitrogen; IFFCO, India | https://nanourea.in/ |
| 6.0 | IFFCO Nano-DAP | 8.0% N and 16% P; IFFCO, India | https://nanodap.in/en |
| 7.0 | Gromor Nano-DAP | 2.0% N and 5.0% $P_2O_5$; Coromandel fertilizers | https://www.coromandel.biz/press-release/nano-dap/ |

simultaneously provides nitrogen and phosphorous to plants (https://nanodap.in/en). Nano-DAP is another fertilizer manufactured by coromandel fertilizers for filar applications (https://www.coromandel.biz/press-release/nano-dap/).

## 4.3 Conclusions

Nanofertilizers recently emerged as one of the promising tools for achieving global food security and agricultural sustainability, thus making it an advantageous option over conventional chemical fertilizers. Nanofertilizers enable slow release of nutrients, targeted delivery of nutrients, plant growth stimulation, and reduction in the loss of water and nutrients. As the global population is expected to cross the nine billion mark by 2050, the development and application of nanofertilizers become of primary importance to meet the increased food demand and thereby reduce unwanted environmental impacts. The major advantages of using nanofertilizers include boosting of agricultural production through improved nutrient use efficiencies and reduction in runoff losses and other environmental impacts. Nanofertilizers improve nutrient uptake by plants and water use efficiency, leading to better crop production and quality. Moreover, nanofertilizers contribute to the sustainability of agriculture by lowering the need for fertilizers and other inputs to achieve the same yield, thus improving crop quality. Additionally, it helps to mitigate nutrient loss through leaching and volatilization. Regular monitoring and assessment of long-term impacts of nanofertilizers on crop yield, soil and water quality, and environmental sustainability are necessary to ensure maximization of their benefits while minimizing any negative consequences. Nanofertilizers should be mainstreamed into the modern agricultural practices to achieve agricultural sustainability, thereby ensuring global food security and minimizing the environmental footprints.

## References


Adam D (2021) How far will global population rise? Researchers can't agree. Nature 597(7877):462–465

Adhikari T, Ramana S (2019) Nano fertilizer: its impact on crop growth and soil health. J Res Pjtsau 47:1–70

Afify RR, El-Nwehy SS, Bakry AB, Abd El-Aziz ME (2019) Response of peanut (Arachis hypogaea L.) crop grown on newly reclaimed sandy soil to foliar application of potassium nanofertilizer. Middle East J Appl Sci 9(1):78–85

Aguilar-Pérez KM, Avilés-Castrillo JI, Ruiz-Pulido G (2020) Nano-sorbent materials for pharmaceutical-based wastewater effluents-An overview. Case Stud Chem Environ Eng 2:100028

Alexandratos N, Bruinsma J (2012) World agriculture towards 2030/2050: the 2012 revision. ESA Working Paper No. 12-03. FAO, Rome

Arora NK (2018) Agricultural sustainability and food security. Environ Sustain 1:217–219

Babu S, Singh R, Yadav D, Rathore SS, Raj R, Avasthe R, Yadav SK, Das A, Yadav V, Yadav B, Shekhawat K (2022) Nanofertilizers for agricultural and environmental sustainability. Chemosphere 292:133451

Bilal M, Barceló D, Iqbal HM (2021) Occurrence, environmental fate, ecological issues, and redefining of endocrine disruptive estrogens in water resources. Sci Total Environ 800:149635

Bose P (2020) Uses of nanotechnology in fertilizers. AZoNano, Sydney. https://www.azonano.com/article.aspx?ArticleID=5446

Brownlie WJ, Sutton MA, Cordell D, Reay DS, Heal KV, Withers PJ, Vanderbeck I, Spears BM (2023) Phosphorus price spikes: a wake-up call for phosphorus resilience. Front Sustain Food Syst 7:1088776

Cheng Y, Wang J, Zhang JB, Müller C, Wang SQ (2015) Mechanistic insights into the effects of N fertilizer application on N 2 O-emission pathways in acidic soil of a tea plantation. Plant Soil 389:45–57

Chugh G, Siddique KH, Solaiman ZM (2021) Nanobiotechnology for agriculture: smart technology for combating nutrient deficiencies with nanotoxicity challenges. Sustain For 13(4):1781

Das A, Babu S, Yadav GS, Ansari MA, Singh R, Baishya LK et al (2016) Status and strategies for pulses production for food and nutritional security in north-eastern region of India. Indian J Agron 61:43–57

Delfani M, Baradarn Firouzabadi M, Farrokhi N, Makarian H (2014) Some physiological responses of black-eyed pea to iron and magnesium nanofertilizers. Commun Soil Sci Plant Anal 45(4):530–540

Deo HR, Chandrakar T, Srivastava LK, Nag NK, Singh DP, Thakur A (2022) Effect of Nano-DAP on yield, nutrient uptake and nutrient use efficiency by rice under Bastar plateau. Pharma Innovat J 11(9):1463–1465

DeRosa MC, Monreal C, Schnitzer M, Walsh R, Sultan Y (2010) Nanotechnology in fertilizers. Nat Nanotechnol 5(2):91–91

Deshpande P, Dapkekar A, Oak MD, Paknikar KM, Rajwade JM (2017) Zinc complexed chitosan/TPP nanoparticles: a promising micronutrient nanocarrier suited for foliar application. Carbohydr Polym 165:394–401

Dirzo R, Raven PH (2003) Global state of biodiversity and loss. Annu Rev Environ Resour 28(1):137–167

Du W, Yang J, Peng Q, Linag X, Mao H (2019) Comparison study of zinc sulphate on wheat growth: from toxicity and zinc biofortification. Chemosphere 27:109–116

Ekinci M, Dursun A, Yildirim E, Parlakova F (2014) Effects of nanotechnology liquid fertilizers on the plant growth and yield of cucumber (Cucumis sativus L.). Acta Scientiarum Polonorum. Hortorum Cultus 13(3)

Espevig T, Pettersen TO, Aamlid TS (2020) Potential of NanoPro to reduce fungicide rate for control of Microdochium nivale on an annual bluegrass (Poa annua) green. In: BIO Web of Conferences, vol 18. EDP Sciences, p 00007

Ganesan V (2015) Biogenic synthesis and characterization of selenium nanoparticles using the flower of Bougainvillea spectabilis Willd. Int J Sci Res 4:690–695

Ghafariyan MH, Malakouti MJ, Dadpour MR, Stroeve P, Mahmoudi M (2013) Effects of magnetite nanoparticles on soybean chlorophyll. Environ Sci Technol 47(18):10645–10652

Ghormade V, Deshpande MV, Paknikar KM (2011) Perspectives for nano-biotechnology enabled protection and nutrition of plants. Biotechnol Adv 29(6):792–803

Hoagland DR, Arnon DI (1950) The water-culture method for growing plants without soil. California agricultural experiment station. Circular 347(2nd edition)

Jha A, Pathania D, Damathia B, Raizada P, Rustagi S et al (2023) Panorama of biogenic nano fertilizers: a road to sustainable agriculture. Environ Res 235:116456

Kabata-Pendias A, Pendias H (1984) Trace elements in soils and plants. CRC Press, Boca Raton, FL

Khot LR, Sankaran S, Maja JM, Ehsani R, Schuster EW (2012) Applications of nanomaterials in agricultural production and crop protection: a review. Crop Prot 35:64–70

Komarek AM, Drogue S, Chenoune R, Hawkins J, Msangi S, Belhouchette H, Flichman G (2017) Agricultural household effects of fertilizer price changes for smallholder farmers in Central Malawi. Agric Syst 154:168–178

Kopittke PM, Menzies NW, Wang P, McKenna BA, Lombi E (2019) Soil and the intensification of agriculture for global food security. Environ Int 132:105078

Kumar Y, Singh T, Raliya R, Tiwari KN (2021) Nano fertilizers for sustainable crop production, higher nutrient use efficiency and enhanced profitability. Indian J Fertilizer 17:1206–1214

Kumar A, Singh K, Verma P, Singh O, Panwar A, Singh T, Kumar Y, Raliya R (2022) Effect of nitrogen and zinc nanofertilizer with the organic farming practices on cereal and oil seed crops. Sci Rep 12(1):6938

Kyriakou V, Garagounis I, Vourros A, Vasileiou E, Stoukides M (2020) An electrochemical Haber-Bosch process. Joule 4(1):142–158

Lal R (2008) Promise and limitations of soils to minimize climate change. J Soil Water Conserv 63(4):113A–118A

Lee WM, An YJ, Yoon H, Kweon HS (2008) Toxicity and bioavailability of copper nanoparticles to the terrestrial plants mung bean (Phaseolus radiatus) and wheat (Triticum aestivum): plant agar test for water-insoluble nanoparticles. Environ Toxicol Chem 27(9):1915–1921

Lin D, Xing B (2008) Root uptake and phytotoxicity of ZnO nanoparticles. Environ Sci Technol 42(15):5580–5585

Liu R, Lal R (2014) Synthetic apatite nanoparticles as a phosphorus fertilizer for soybean (Glycine max). Sci Rep 4(1):5686

Liu R, Lal R (2015) Potentials of engineered nanoparticles as fertilizers for increasing agronomic productions. Sci Total Environ 514:131–139

Liu XM, Zhang FD, Zhang SQ, He XS, Wang RF, Feng ZB, Wang YJ (2005) Responses of peanut to nano-calcium carbonate. J Plant Nutrit Fertilizer 11(3):385–389

Mahajan P, Dhoke SK, Khanna AS (2011) Effect of nano-ZnO particle suspension on growth of mung (Vigna radiata) and gram (Cicer arietinum) seedlings using plant agar method. J Nanotechnol 2011:1

Mahapatra DM, Satapathy KH, Panda B (2022) Biofertilizers and nanofertilizers for sustainable agriculture: phycoprospects and challenges. Sci Total Environ 803:149990

Manikandan A, Subramanian K (2016) Evaluation of zeolite-based nitrogen nano-fertilizers on maize growth, yield and quality on inceptisols and alfisols. Int J Plant Soil Sci 9:1–9

Manjunatha SB, Birader D, Aladakatti Y (2016) Nanotechnology and its applications: a review. J Farm Sci 29:1–13

Menossi M, Casalongué C, Alvarez VA (2022) Bio-nanocomposites for modern agricultural applications. Handbook of consumer nanoproducts. Springer, pp 1201–1237

Molina-Santiago C, Matilla MA (2020) Chemical fertilization: a short-term solution for plant productivity? Microb Biotechnol 13(5):1311

Musante C, White JC (2012) Toxicity of silver and copper to Cucurbita pepo: differential effects of nano and bulk-size particles. Environ Toxicol 27(9):510–517

Nair R, Varghese SH, Nair BG, Maekawa T, Yoshida Y, Kumar DS (2010) Nanoparticulate material delivery to plants. Plant Sci 179(3):154–163

Nekrasova GF, Ushakova OS, Ermakov AE, Uimin MA, Byzov IV (2011) Effects of copper (II) ions and copper oxide nanoparticles on Elodea densa Planch. Russ J Ecol 42:458–463

Oancea F (2022) Development of safe Nanoagrochemicals—the Nanoporous route. Chem Proc 7(1):68

Okey-Onyesolu CF, Hassanisaadi M, Bilal M, Barani M, Rahdar A, Iqbal J, Kyzas GZ (2021) Nanomaterials as nanofertilizers and nanopesticides: An overview. ChemistrySelect 6(33):8645–8663

Owuor PO, Othieno CO, Kamau DM, Wanyoko JK (2011) Effects of long-term fertilizer use on a high-yielding tea clone AHPS15/10: soil pH, mature leaf nitrogen, mature leaf and soil phosphorus and potassium. Int J Tea Sci 8(1):15–51

Perrin TS, Drost DT, Boettinger JL, Norton JM (1998) Ammonium-loaded clinoptilolite: a slow-release nitrogen fertilizer for sweet corn. J Plant Nutr 21(3):515–530

Pettersen T, Aamlid T (2019) Testing of Nano-Gro on established turf at fairway mowing height (15 mm). NIBIO Rapport

Qamar SA, Qamar M, Basharat A, Bilal M, Cheng H, Iqbal HM (2022) Alginate-based nano-adsorbent materials–bioinspired solution to mitigate hazardous environmental pollutants. Chemosphere 288:132618

Raliya R, Saharan V, Dimkpa C, Biswas P (2017a) Nanofertilizer for precision and sustainable agriculture: culture state and future perspectives. J Agric Food Chem 66(26):6487–6503

Raliya R, Saharan V, Dimkpa C, Biswas P (2017b) Nanofertilizer for precision and sustainable agriculture: current state and future perspectives. J Agric Food Chem 66(26):6487–6503

Rathore SS, Shekhawat K, Singh RK, Upadhyay PK, Shekhawat R, Premi OP (2019) Effect of nano-particles on growth, productivity, profitability of Indian mustard (*Brassica juncea*) under semi-arid conditions. Indian J Agric Sci 89:1145–1150

Rathore R, Hasan A, David AA, Thomas T, Reddy IS (2022) Effect of different levels of nano urea and conventional fertilizer on soil health of maize (*Zea mays* L.) Var, P3544 in an Inceptisols of Prayagraj, (U.P) India. Pharma Innovat J 11(8):560–563

Rosa L, Gabrielli P (2022) Energy and food security implications of transitioning synthetic nitrogen fertilizers to net-zero emissions. Environ Res Lett 18(1):014008

Saharan V, Kumaraswamy RV, Choudhary RC, Kumari S, Pal A, Raliya R, Biswas P (2016) Cu-chitosan nanoparticle mediated sustainable approach to enhance seedling growth in maize by mobilizing reserved food. J Agric Food Chem 64(31):6148–6155

Satyanarayana T, Reddy SS (2018) A review on chemical and physical synthesis methods of nanomaterials. Int J Res Appl Sci Eng Technol 6(1):2885–2889

Schnitkey G, Zulauf C, Swanson K, Paulson N (2021) 2022 planting decisions, nitrogen fertilizer prices, and corn and soybean prices. farmdoc daily 11(145)

Shah V, Belozerova I (2009) Influence of metal nanoparticles on the soil microbial community and germination of lettuce seeds. Water Air Soil Pollut 197:143–148

Shang Y, Hasan MK, Ahammed GJ, Li M, Yin H, Zhou J (2019) Applications of nanotechnology in plant growth and crop protection: a review. Molecules 24(14):2558

Sharma G, Sharma AR, Bhavesh R, Park J, Ganbold B, Nam JS, Lee SS (2014) Biomolecule-mediated synthesis of selenium nanoparticles using dried Vitis vinifera (raisin) extract. Molecules 19(3):2761–2770

Singh NRR, Sarma SS, Rao TN, Pant H, Srikanth VVSS, Kumar R (2021) Cryo-milled nano-DAP for enhanced growth of monocot and dicot plants. Nanoscale Adv 3(16):4834–4842

Song G, Gao Y, Wu H, Hou W, Zhang C, Ma H (2012) Physiological effect of anatase $TiO_2$ nanoparticles on Lemna minor. Environ Toxicol Chem 31(9):2147–2152

Subramani T, Velmurugan A, Bommayasamy N, Swarnam TP, Ramakrishna Y, Jaisankar I, Singh L (2023) Effect of nano urea on growth, yield and nutrient use efficiency of okra under tropical Island ecosystem. Int J Agric Sci 9:134–139

Tarafdar JC, Raliya R, Rathore I (2012) Microbial synthesis of phosphorous nanoparticle from tri-calcium phosphate using Aspergillus tubingensis TFR-5. J Bionanosci 6(2):84–89

Taran NY, Gonchar OM, Lopatko KG, Batsmanova LM, Patyka MV, Volkogon MV (2014) The effect of colloidal solution of molybdenum nanoparticles on the microbial composition in rhizosphere of Cicer arietinum L. Nanoscale Res Lett 9:1–8

Tiwari KN, Kumar Y, Nayak RK, Rai A, Singh J, Srivastava SSA (2021) Nano-urea for enhancing yield and farmers profit with potato in Uttar Pradesh. Ann Plant Soil Res 23:495–500

Upadhyay PK, Dey A, Singh VK, Dwivedi BS, Sngh T, Rajanna GA et al (2023a) Conjoint application of nano-urea with conventional fertilizers: an energy efficient and environmentally robust approach for sustainable crop production. PLoS One 18:e0284009

Upadhyay PK, Singh VK, Rajanna GA, Dwivedi BS, Dey A, Singh RK, Rathore SS, Shekhawat K, Babu S, Singh T, Kumar Y, Singh C, Rangot M, Kumar A, Sarkar S, Dash S, Rawat S (2023b) Unveiling the combined effect of nano fertilizers and conventional fertilizers on crop productivity, profitability and soil well-being. Front Sustain Food Syst 7. https://doi.org/10.3389/fsufs.2023.1260178

Van Dijk M, Morley T, Rau ML, Saghai Y (2021) A meta-analysis of projected global food demand and population at risk of hunger for the period 2010–2050. Nature Food 2(7):494–501

Vitousek PM, Aber JD, Howarth RW, Likens GE, Matson PA, Schindler DW, Schlesinger WH, Tilman DG (1997) Human alteration of the global nitrogen cycle: sources and consequences. Ecol Appl 7(3):737–750

Yadav DN, Kumar R, Verma AK, Kumar P (2021a) Effect of foliar application of n Meena, R.H., G. Jat and D. Jain (2021) impact of foliar application of different nano-fertilizers on soil microbial properties and yield of wheat. J Environ Biol 42:302–308

Yadav DN, Kumar R, Verma AK, Kumar P (2021b) Effect of foliar application of nano-fertilizers on soil health and productivity in transplanted rice (*Oryza sativa* L.). Pharma Innovat J 10(12):1263–1265

Yang F, Liu C, Gao F, Su M, Wu X, Zheng L, Hong F, Yang P (2007) The improvement of spinach growth by nano-anatase $TiO_2$ treatment is related to nitrogen photoreduction. Biol Trace Elem Res 119:77–88

Yusuf M, Fariduddin Q, Hayat S, Ahmad A (2011) Nickel: an overview of uptake, essentiality and toxicity in plants. Bull Environ Contam Toxicol 86:1–17

Zotikova AP, Astafurova TP, Burenina AA, Suchkova SA, Morgalev YN (2018) Morphophysiological features of wheat (Triticum aestivum L.) seedlings upon exposure to nickel nanoparticles. Sel'skokhozyaistvennaya Biologiya 53:578–586

# Chapter 5
# Impact of Nanofertilizers on Soil, Plant, and Animal Health

**Avijit Ghosh, Anand Singh, Ashok K. Patra, Amit K. Singh, and Sukanya Misra**

**Abstract** In a startling tale of agricultural excess, the once-fertile soil has suffered greatly from the constant abuse of artificial fertilizers in contemporary farming techniques, which has ruthlessly depleted its essential nutrients and irreversibly damaged the harmony with the surrounding environment. Contrarily, nanofertilizers are more cost-effective and specific than conventional fertilizers because they regulate the release of nutrients through the use of nanoparticles. One or more plant nutrients are included in nanoparticles of nanofertilizers, with at least half of the particles having a diameter of less than 100 nm. They ensure that nutrients are released progressively over an extended period of time, providing plants with a consistent supply of essential components. Furthermore, nanofertilizers can improve environmental sustainability by reducing nutrient leakage and discharge into the environment. Additionally, they can increase the efficiency with which fertilizer is used, which will increase crop yields and lower the total cost of applying fertilizers. Particularly useful are nanofertilizers in regions where conventional fertilizers are inefficient or useless. By lowering the environmental impact of fertilizer application, nanofertilizers can offer a more economical and effective method of fertilizing crops. They are the result of an exciting new technology that has the potential to both increase agricultural sustainability and assist in fulfilling a growing food need. Because they contain nanoparticles, nanofertilizers have some drawbacks at the moment, such as increased manufacturing costs and possible safety and environ-

A. Ghosh (✉) · A. K. Singh
ICAR-Indian Grassland and Fodder Research Institute, Jhansi, Uttar Pradesh, India
e-mail: avijit.ghosh@icar.gov.in

A. Singh
Acharya Narendra Deva University of Agriculture and Technology, Ayodhya, Uttar Pradesh, India

A. K. Patra
ICAR-Indian Institute of Soil Science, Bhopal, Madhya Pradesh, India

S. Misra
Rani Lakshmi Bai Central Agricultural University, Jhansi, Uttar Pradesh, India

 
A. K. Patra (ed.), *Nanofertilizers in Agriculture*, Sustainability Sciences in Asia and Africa, https://doi.org/10.1007/978-981-97-8112-6_5

mental issues. More study is required to completely comprehend the long-term consequences of nanofertilizers on soil health, crop development, and the environment.

**Keywords** Nanofertilizers · Environmental pathways · Ecosystem health · Sustainable agriculture · Toxicity

## 5.1 Introduction

Agriculture has been the back bone for supporting the world's expanding population and propelling the booming economy. In the process, it has become clear that using fertilizers regularly is crucial to maintaining soil fertility and increasing crop yields. Traditional fertilizers are commonly used to support the inherent soil fertility. These fertilizers include urea, nitrogen, potassium, phosphorous, and di ammonium phosphate. However, leaching causes conventional fertilizers to have low nutrient use efficiency (NUE), which results in significant financial losses and diminished soil fertility. Soil fertility has significantly decreased as a result of these nutrients seeping from the soil. This is mostly because conventional fertilizers have a relatively low NUE of about 30%–35% for nitrogen and 15%–20% for phosphorus (Preetha and Balakrishnan 2017). Scientists are already working on creating chemical fertilizers with a slow-release rate. For instance, by mixing urea and hydroxyapatite, they have created a fertilizer that releases plant nutrients gradually (Xiong et al. 2018). Furthermore, the release of excess nutrients has an adverse effect on the environment, which has prompted the emergence of more environmentally friendly and effective fertilizers. In order to guarantee optimal uptake and use in plants, nano-fertilizers—which possess suitable formulations and delivery mechanisms—are crucial in contemporary agriculture (Mittal et al. 2020; Adisa et al. 2019; Fraceto et al. 2016). These nanoscale fertilizers can enhance nutrient usage efficiency and environmental quality by reducing nutrient losses due to leaching and averting chemical modifications by investigating nanoparticles (NPs) based on various metals and metal oxides for agricultural applications (Saharan et al. 2016).

Nanoparticles (NPs) are small molecules, typically between one and one hundred nm in size, possessing physiochemical properties different from bulk materials (El-Saadony et al. 2019, 2020, 2021; Reda et al. 2020, 2021). The NPs' superior properties and functions in terms of all three soil properties are due to their higher surface area to volume ratio. Certain nutrients are provided in a nano form via nanofertilizers, which improve plant development and yield (Dimkpa and Bindraban 2016).Three types of nanofertilizers are distinguished based on the nutrient supply to plants: macro, micro, and nanoparticulate fertilizers (Chhipa and Joshi 2016). Nanofertilizers can disperse in the form of a particle or liquid with a diameter of less than 100 nm (Josef and Katarína 2015). They supply nutrients for crops in a form that is readily absorbed by them, enhancing plant uptake and yield. The pertinent aspects of the nanofertilizers are summarized in Fig. 5.1: (1) using foliar and soil treatments to improve plant growth by providing the necessary nutrients, (2) they

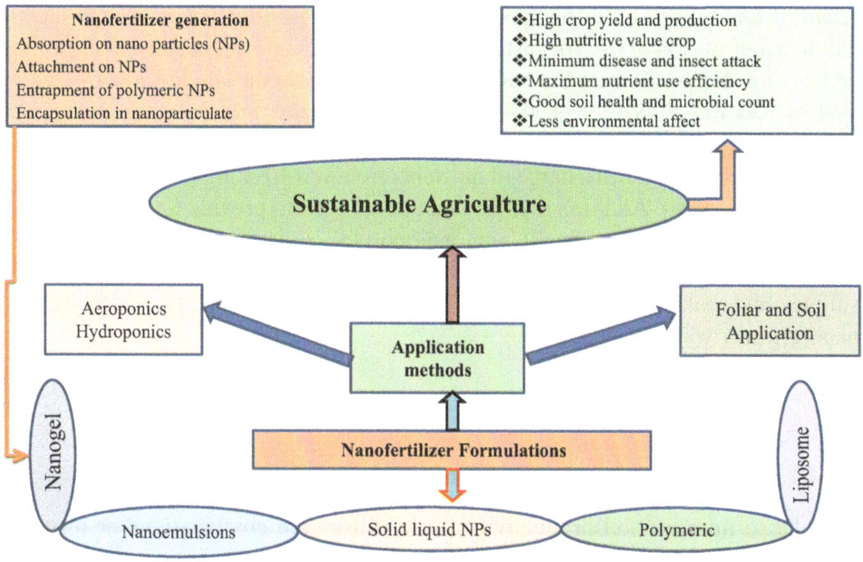


**Fig. 5.1** Nanofertilizers in agriculture. *NPs* nanoparticles, *NFs* nanofertilizers

are a sustainable and affordable source of nutrients, (3) high use efficiency, and (4) important to overcome the pollution problem. Furthermore, nanofertilizers—also referred to as novel fertilizer alternatives—help remove pollutants from water (Guru et al. 2015).

## 5.2 Nanofertilizers

Since nanotechnology has significantly revolutionized current science and is expanding exponentially, it has become increasingly important in the modern day, along with its related applications (Bhardwaj et al. 2022). Nanomaterials range between 1 and 100 nanometer in size.

The unique characteristics of nanomaterials when paired with native and customary practices could lead to a plethora of inventive uses in a range of scientific fields, including agriculture, which needs creative approaches to guarantee global food security (Raliya et al. 2018a, b). Fertilizer use significantly boosted grain development, but poor fertilizer response ratios, uneven fertilization, insufficient organic matter, and intensified multi-micronutrient deficits nationwide caused yields of some crops to stall (Morales-Díaz et al. 2017). Unbalanced fertilization is among the most important issues that soil scientists must take into account while managing nitrogen, among other obstacles.

According to Komarneni (2009), nanotechnology has promise for creating fertilizers with a gradual release. Clays are instrumental in determining the fertility of the soil due to their extensive surface area and adsorptive sites, which facilitate the retention and release of nutrients. One gram of montmorillonite can be stretched to a dimension of 30–40 $m^2$, for instance, if the same clay is reduced by ball milling and its dimension is lowered to nanometers that increase the surface area by several fold of 750 $m^2g^{-1}$ (Macht et al. 2011). Nanofertilizers and nanocomposites can enhance nutrient use efficiency and prevent nutrient fixation or loss to the environment by controlling the discharge of nutrients from fertilizer granules (Subramanian and Tarafdar 2009). Additionally, they can be utilized to provide a variety of nutrients in appropriate ratios. Zeolite and nanoporous zeolite are used as slow-release fertilizers in agriculture. For a period of 60 days, the availability of urea, potassium sulfate, and calcium hydroxyapatite combined with zeolite as a slow-release nanofertilizer was enhanced.

## 5.3 Nanobiofertilizers

In order to improve soil productivity, biofertilizers are combinations or formulations that contain one or more microorganisms that fix atmospheric nitrogen, solubilize phosphorus, or stimulate plant development by generating chemicals that aid in plant growth. (Kole et al. 2013; Malusa and Vassilev 2014; Singh et al. 2016). Consequently, the term "nanobiofertilizer" could refer to a blend of biofertilizers and nanoparticles that are intended to improve plant growth and development (Simarmata et al. 2016). The management of biofertilizer application rates as well as formulation shelf life is critical to the attainment of this objective.

The interaction between microorganisms and nanoparticles, as well as the shelf life and distribution of biofertilizers, are some of the most crucial factors in the formation of nanobiofertilizers. Positive effects of the interaction between gold nanoparticles and plant growth-promoting rhizobacteria have been demonstrated (Malusá et al. 2012; Shukla et al. 2015; Duhan et al. 2017). Conversely because of interference of silver nanoparticles in biological processes in microorganisms, such as the structure and functions of cell membranes, they cannot be utilized together with biofertilizers (Duhan et al. 2017). However, the inclusion of nanoparticles can extend the shelf life of biofertilizers, which is mostly a limiting factor in these formulations. By using nanoformulations, biofertilizer stability against desiccation, heat, and UV inactivation can be improved. For instance, desiccation-resistant formulations can be created using polymeric nanoparticle coatings, extending the products' useful lives (Simarmata et al. 2016; Shukla et al. 2015; Duhan et al. 2017; Jampílek and Kralova 2017). Additionally, biofertilizers can be delivered to soil and plants more effectively by using nanoparticles. Experiments including hydrophobic silica NPs in the water-in-oil emulsion have demonstrated improved product distribution and increased shelf life due to less desiccation (Kaushik and Djiwanti 2017). As a result, nanobiofertilizers can address some of the drawbacks of biofertilizers, but further study and development of this technology are still needed.

## 5.4 Effect of Nanofertilization on Soil Microorganism

Studies on how soil microflora is affected by nanofertilizers in terms of viability and susceptibility are being conducted by certain researchers. According to one study (Rajput et al. 2018), nanofertilization improved the ecological habitat, increased the amount of soil microorganisms, and increased soil nutrients. In addition, there were noticeably more soil microorganisms treated with nanofertilizer than those treated with conventional fertilizers. Humic acid (HA) is generated in enormous quantities via the delayed release of nanofertilizers. The foundation of soil fertility is humic acid, which gives soil microorganisms supplies of carbon and nitrogen (VandeVoort and Arai 2019). Furthermore, HA can control soil pH, stimulate microbial growth and reproduction, raise the amount and variety of soil microorganisms, and direct or indirect enhancement of soil temperature, moisture content, and gas permeability (Subramanian and Tarafdar 2009).

The researchers showed that high concentrations of titanium oxide (TiO2) can upset the bacterial-clover symbiosis and impede clover growth in an additional study on the effects of nanoparticles on soil microorganisms and crops in liquid cultures (Morales-Díaz et al. 2017). The assimilation of titanium dioxide by plants did not increase, and the mobility of the NPs in natural soil was found to be inadequate. Another study examined the effects of organic nano NPK formulations on the microbiological load (fungi and bacteria) and the enzymatic activity (urease, acid phosphatase, dehydrogenase, and alkaline phosphatase) at the experimental locations. The formulations were applied to the soil and foliage. The particle sizes of liquid organic nano NPK and granular organic nano NPK were determined to be 67.30 nm and 89.26 nm, respectively, after fertilizer sample analysis (Nibin et al. 2019). Soil samples were collected during the last crop harvest to assess the microbial load and enzymatic activity of the soil. Based on the findings, it was concluded that the highest post-harvest soil conditions in terms of bacterial count, dehydrogenase, urease, and acid phosphatase content were achieved by applying farm yard manure (FYM) (12 t $ha^{-1}$), soil application of nano NPK (12.5 kg $ha^{-1}$), and foliar application of nano NPK (0.4%) (Nibin et al. 2019).

## 5.5 Utilizing Nanoparticles to Oppose Plant Pathogens

Numerous studies have used silver nanoparticles to remove harmful bacteria from plants. It has been demonstrated that white radish (*Raphanussativus var. aegyptiacus*) may be biologically synthesized into silver nanoparticles (NPs) with sizes ranging between 6 and 38 nm. The activity and viability of land snails, as well as the frequency of fungal communities in the surrounding soil, were reduced as a consequence of the exposure of the snails and soil matrix to silver nanoparticles in a laboratory experiment. (20% of the snails treated with silver nanoparticles perished) (Ali et al. 2015). Through testing, *Serratia* sp. BHU-S4 culture supernatant was used to form spherical-shaped silver nanoparticles with a size range of around 10–20 nm. These nanoparticles have shown promise in the treatment of wheat spot blotch disease. Silver nanoparticles significantly reduced the wheat spot blotch pathogen, *Bipolarissorokiniana* (Mishra et al. 2014).

## 5.6 Using Nano-Biofertilizers to Encourage Plant Growth

According to research, using silver and gold NP materials to promote growth has also been successful (Dikshit et al. 2013). Under in vitro settings, these NPs containing natural biofertilizers such as *Paeni bacilluselgii, Bacillus subtilis*, and *Pseudomonas fluorescens* have demonstrated very good growth enhancement. Nano-biofertilizers are required in extremely small quantities in comparison to other fertilizers, as they can fertilize several hectares of crops with a single liter. Consequently, their costs are manageable. The rhizosphere zone is home to a variety of soil microorganisms, but plant growth-promoting rhizobacteria (PGPR) are the greatest at encouraging plant growth. It was determined how gold nanoparticles affected *Pseudomonas fluorescens*, *Bacillus subtilis*, *Paenibacillus elgii*, and *Pseudomonas putida*. *Pseudomonas putida* showed no discernible response to gold nanoparticles, either favorable or negative. Gold nanoparticles can be used as nano-bio-fertilizers since a significant rise in *P. fluorescens*, *P. elgii*, and *B. subtilis* was noted (Shukla et al. 2015).

## 5.7 Nanofertilizers Enhance Crop Yield and Nutrient Use Efficiency (NUE)

Both foliar and direct application of nanofertilizers promotes plant development (Table 5.1). The most effective solution for preventing eutrophication and thereby increasing the efficacy of nutrient use in agriculture may be nanofertilizers (Godfray et al. 2010; Shukla et al. 2019; Jiang et al. 2021). Furthermore, studies have revealed that the application of "controlled release fertilizer" increases wheat yield and soil residual mineral nitrogen by 6% and 10%, respectively, while reducing nitrogen leaching and runoff loss by 25% and 22%, respectively, in comparison to those achieved with traditional fertilizers (Jiang et al. 2021; Liu et al. 2016; Kah et al. 2014). The availability of micronutrients during crop growth is detrimental, as it results in fruits and vegetables that are low in nutritional value (López-Valdez et al. 2018). According to López-Valdez et al. (2018), NPs are an efficient way to supply crops and fresh food products with micronutrient deficiencies like zinc and iron. When sprayed at a specified concentration for a variety of crops, several macro- and micronutrient-based nanofertilizers have demonstrated remarkable targeted delivery (Liu and Lal 2014). Nanofertilizers are already widely recognized for their

**Table 5.1** Impact of nanoparticles/nanofertilizers on crops under unfavorable environmental variables

| NPs | Plant | Application type | Concentration range | Impacts | Source |
|---|---|---|---|---|---|
| NPK NPs | Wheat | Foliar spray | 500, 60, and 400 ppm | Application after the date of planting within 21 days. A notable rise in wheat grains' overall saccharide content | Abdel-Aziz et al. (2018) |
| Zn, Fe and NPK | Chickpea | Foliar application | 20 L/plot at each stage | 1st spray at 4–6 leaf stage, second spray at 30 days, and third spray during pod filling. There was a notable increase in both biological and seed yield | Drostkar et al. (2016) |
| N, P and NPK NPs | Wheat | Foliar application | | Significant variations in the length of the shoots, roots, and other plant growth factors | Al-Juthery et al. (2019) |
| nCu, nFe, nCo (metal NPs) | *Zea mays L.* | Soil irrigation | 3–5 ppm | Improved the frequency, duration, early growth, enzymatic activity, and metabolism of SOD in plant leaves to increase the capacity of the plants to withstand stress | Hoang et al. (2019) |
| nFe2O3 | *Solanum lycopersicum L.* | Hydroponic | 50–800 ppm | Superior seed germination, morphological characteristics, dry weight, and uptake of Fe relative to typical plants | Shankramma et al. (2016) |
| nFeS | *Brassica juncea L.* | Foliar | 2–10 ppm | Increased enzymatic activity, membrane damage, photosynthetic pigments, improved agronomic features, and nutrient absorption | Rawat et al. (2017) |
| nFe | *Capsicum annuum L.* | Foliar | 0.002–2 mM $L^{-1}$ | It has been observed that low doses of nFe positively impact plant growth and development. Improved grana stacking and chloroplast functional capability. Elevated concentrations of FeNPs have been shown to be detrimental to plants and may even halt the flow of Fe nutrients | Yuan et al. (2018) |

(continued)

**Table 5.1** (continued)

| NPs | Plant | Application type | Concentration range | Impacts | Source |
|---|---|---|---|---|---|
| nZnO | *Zea mays L.* | Foliar | 150–300 ppm | It has been observed that low doses of nFe positively impact plant growth and development. Improved grana stacking and chloroplast functional capability. Elevated concentrations of FeNPs have been shown to be detrimental to plants and may even halt the flow of Fe nutrients | Raddy et al. (2018) |
| | *Solanum lycopersicum Mill.* | Tissue culture | 15–30 ppm | Lower doses of ZnO NPs proved more suitable than higher doses. Different cultivars showed varying levels of stress tolerance | Alharby et al. (2016) |
| | *Zea mays L.* | Foliar | 50–2000 ppm | Enhanced Zn accumulation in grains, biomass, productivity, seed germination rate, and seedling vigor index | Subbaiah et al. (2016) |
| | *Arachishypogaea L.* | Soil | 100–500 ppm | The NP treatment had a favorable impact on total phenols, total soluble sugar, reducing sugar, and photosynthetic pigments. Biochemical, morphological and yield traits improved | Rajiv and Vanathi (2018) |
| | *Sorghum bicolor L.* | Soil and foliar | 6 mg $kg^{-1}$ soil | Increased grain nutritional profile, NUE, uptake of N and K elements, and enhanced plant performance and yield component when compared to normal plants | Dimkpa et al. (2017) |
| $Mn_3O_4$ | *Cucumis sativus L.* | Foliar | 1–5 mg $plant^{-1}$ | Substantially enhanced biomass, photosynthetic responses, chlorophyll concentration, and plant development. Enhanced natural antioxidant defense systems | Lu et al. (2020) |
| nZ (Zein NPs) | *Saccharum* spp. | Hydroponic | 0.88–1.75 mg $mL^{-1}$ | ZNPs are absorbed in considerable quantities by the roots of the sugarcane plant, and Z particles are found in both the endodermis and the epidermis of the roots. Increased absorption of nutrients by the plant system | Prasad et al. (2017) |

larger surface areas, which are a result of their smaller particulates. This provides the plant system with additional surface area to support a variety of metabolic processes, which resulted in the synthesis of additional photosynthates. They are highly soluble in several solvents, including water, and have significant reactivity with other chemicals. Nanofertilizers consist of particles that are less than 100 nm in size, which enables the nutrient to penetrate the plant more effectively from surfaces such as soil or foliage. This enhances the nutritional quality parameters of the plant by expediting the reaction or synthesis process in the plant system, which in turn promotes nutrient absorption and penetration (e.g., protein, oil, and sugar content) (Singh et al. 2017). When zinc and iron are applied to plants, the amount of protein, starch, Indole acetic acid (IAA), chlorophyll, and total carbohydrates rises (Rajaie and Ziaeyan 2009). Peanut plant growth and photosynthesis are enhanced by nanoferric oxide. (Liu et al. 2010).

Fertilizers that are encapsulated in nanoparticles will improve the availability and absorption of nutrients by crop plants (Tarafdar et al. 2012). The denitrification, volatilization, leaching, and fixation of nutrients in the soil can be prevented by zeolite-based nano-fertilizers, which progressively release nutrients to the cereal plant. Two examples of such nutrients are nitrate-nitrogen (NO3-N) and ammonium-nitrogen (NH4-N). Under greenhouse circumstances, nano-fertilizer formulations may improve nutrient utilization efficiency. These nano formulations, which can also be in the form of nano-pesticides, have the potential to safeguard crops from fungal and bacterial diseases. However, they may also unintentionally harm non-target plant-associated microorganisms, such as nitrogen-fixing bacteria or mycorrhizal communities, which are essential to plant nutrition. Moreover, this broad-spectrum anti-microbial treatment may have a detrimental effect on the long-term preservation of soil fertility by inhibiting the decomposition of organic matter (OM) by beneficial soil microorganisms. In that case, these treatments with nano-formulations might have negative long-term consequences on soil fertility and plant yields. Although we already know certain adverse effects of nanoparticles on the soil bacterial population, there is still a lack of information regarding the effects of functionalization, concentration, exposure period, and soil texture on the nanoparticle effect expression.

## 5.8 Nanofertilizer (NF) and Soil Health

The application of NFs through soil irrigation assures double advantages, namely soil improvement to maximize plant development productivity, as the application of larger quantities of inorganic fertilizers to farming land may not be available to plants. (Mahapatra et al. 2022). As a result, NFs could be a preferable strategy for the roots' absorption of nutrients. A variety of edaphic parameters regulate the range of mineral elements in the soil, which can also influence microbial colonies and

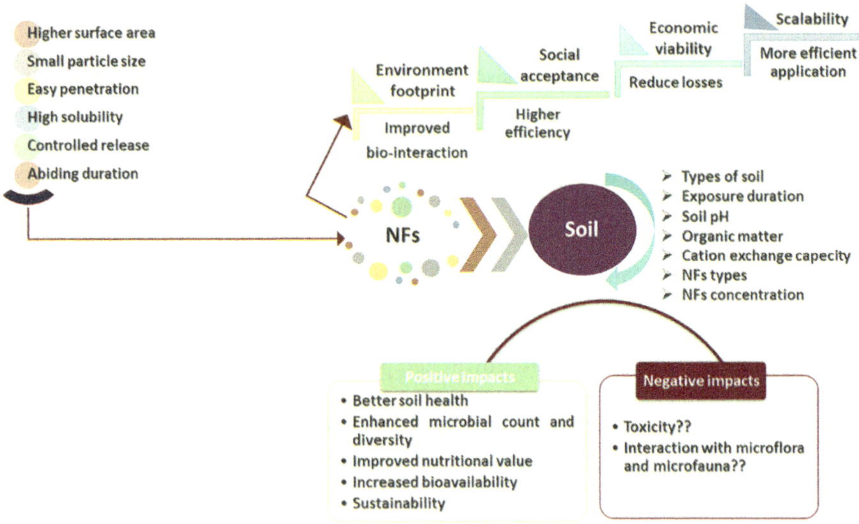


**Fig. 5.2** An overview of the role of nanofertilizers in soil

rhizospheric microbial biomass to enhance soil fertility, water availability, and plant growth (Fig. 5.2).

The development of roots is influenced by soil aeration, nutrient availability, pH, and soil texture. Roots are a critical point of interaction between plants and soil, as they facilitate the absorption of nutrients, water, and other physiologically significant substances.

Diffusion and bulk (mass) flow are the main processes that carry nutrients from the soil to the aerial portions of plant tissues. The bulk flow is defined as the pressure-driven distribution of solutes—water—via xylem, controlled by transpiration and soil nutrient availability. Diffusion, on the other hand, is the movement of minerals from one cell to another along a concentration gradient (Marschner 2011). A variety of absorption and assimilation pathways, such as the aerial surface, roots, granules, interacting atmospheric variables, cell wall rigidity, and the physiological, anatomical, and biochemical activities of the plant species/cultivars, are associated with the accumulation of NPs. NPs effectively regulate nutrient release by exerting a greater surface tension on the surfaces of fertilizer particles than that of conventional fertilizers (Adisa et al. 2019).

The soil microbial community was demonstrated to be unaffected by TiO2 NPs in the 1–100 ppm range, while CuO, ZnO, and Ag NPs were found to be hazardous. Metal nanoparticles (NPs) may penetrate seeds and be translocated into seedlings to facilitate plant development through seed priming (Seleiman et al. 2020). Nutrients in NFs can be progressively released, either independently or in conjunction with organic or synthetic fertilizers. Natural fertilizers may require 40–50 days to release all of their nutrients, whereas commercial fertilizers release nutrients within 4–10 days (Seleiman et al. 2020).

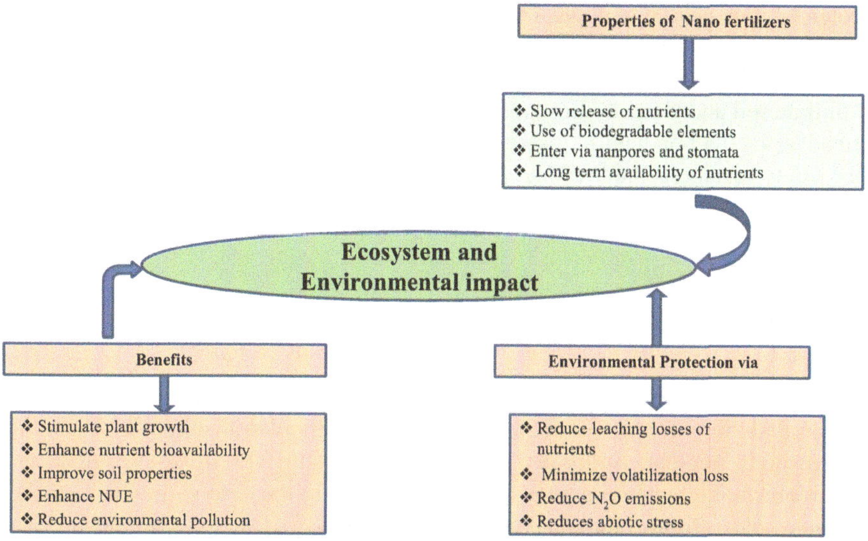


**Fig. 5.3** Influence of nanofertilizers on the ecosystem and environment

## 5.9 Fate of Nanofertilizers on Ecosystems

Since nanofertilizers release some nutrients at a slower rate than other fertilizers, they are thought to be less damaging to crop plants and the environment (Fig. 5.3) (Monreal et al. 2016; Kottegoda et al. 2017). Rain, wind, trophic translocation and contaminated leachate are some of the ways that the nanofertilizers used for plants might find their way from crop fields into the soil, water, and atmosphere. Because the accidental presence of NFs in goods intended for domestic animal or human consumption is a poorly researched topic, the use of NFs should be carefully analyzed. Numerous research on the topic shows that microorganisms in the soil, sediment, and plant roots absorb nanoparticles. These NPs move from one trophic level of food chain to another higher level, implying their negative consequences even on offspring. This phenomenon has been verified to occur in various plant-herbivore-carnivore food chains and has also been observed in marine organisms (Morales-Díaz et al. 2017).

## 5.10 Nanofertilizer and Human Health

Applications of nanotechnology in agriculture have implications for human health, including food safety and environmental preservation. The drawback of nanotechnology is that, because of their small size and bigger surface/mass ratio, nanoparticles (NPs) are more likely to accumulate in different species and have a far greater potential to penetrate biological membranes (Xuan et al. 2023). Agriculture will be greatly impacted by nanotechnology in a number of ways, some of which may be positive and some of which may be harmful to human health. For instance, NPs may infiltrate soil and water bodies, potentially harming ecosystems, and they may have an adverse effect on helpful insects like bees. Consequently, it is crucial to reassess the safety of nanotechnology and NPs in relation to human health and the environment. (Raliya et al. 2018a, b).

## 5.11 Nanofertilizer Toxicity

Despite some associated toxicity hazards, the use of NPs as fertilizers to increase the availability of plant nutrients and increase agricultural output is becoming increasingly popular. Additional potential issues with NPs' toxicity, safety, and environmental effects necessitate investigation due to the ease with which smaller particles can infiltrate biological systems (Li et al. 2016). Due to their inherent characteristics, such as size, shape, charge and surface area, solubility, crystal phases, coatings, material types, and dosage concentrations, numerous nanoparticles are believed to pose potential health risks to humans. In addition, NP behavior, transport, and toxicity may be influenced by a combination of environmental variables, including temperature, pH, ionic strength, salinity, and organic matter. Due to varying reaction settings, soil characteristics, and dosages utilized in the tests, recent research have demonstrated the detrimental impacts of NPs regarding the dynamics of soil organic matter (Schlich and Hund-Rinke 2015; Rahmatpour et al. 2017).

## 5.12 Nanofertilizers Over Conventional Chemical Fertilizers

Compared to conventional fertilizers, nanofertilizers are less hazardous to the environmental and human health (Table 5.2). Furthermore, they maximize profit, minimize expenses, and improve crop quality, yield, and soil fertility. Because chemical fertilizers are synthetic—that is, composed of non-organic agricultural ingredients—farmers utilize them in enormous quantities to increase crop yield. Chemical fertilizers work faster than organic ones because they are less expensive, readily dissolved in water, and available in granular or liquid form. However, P fertilizers do not easily dissolve in water due to the presence of some insoluble compounds such as triple superphosphate, diammonium phosphate, and mono-ammonium phosphate. Soil bacteria decompose organic components, including bulky organic manure, bird droppings, food waste, and sewage sediment, to liberate essential nutrients. This natural fertilizer is more environmentally beneficial due to its ability to improve soil texture, increase bacterial and fungal activity, and extend water retention. The majority of the nutrients that this material produces are N, K, and P, which shield plants from illnesses and pests. The delayed release of nutrients is the

**Table 5.2** A comparative analysis between conventional fertilizers and nanofertilizers (Mushtaq et al. 2020)

| Properties | Nanotechnology | Conventional fertilizers |
|---|---|---|
| Solubility and dispersion | Nanotechnology-formulated micronutrients have the potential to enhance and augment bioavailability by mitigating soil absorption and fixing | Reduced solubility as a result of the macroscale |
| Mode of nutrient release | For water-soluble fertilizers, nutrient release can be regulated in terms of both rate and pattern through encapsulation | Overuse of fertilizers leads to toxicity and an ecological imbalance in the soil |
| Pattern of nutrient release | They could prolong the time that fertilizers effectively feed nutrients to the soil for plant uptake | When they are delivered, plants use them, and the remaining portion ends up in the soil as insoluble salts |
| Uptake efficiency | Due to their nanoscale, they may enhance fertilizer efficiency and nutrient uptake | It reduces effectiveness and prevents the bulk amount from reaching the roots of plants |
| Rate of nutrient loss | They lessen the soil's loss of fertilizer nutrients | Due to leaching, runoff, and drift, they exhibit significant nutrient losses |

primary advantage of this type of fertilizer over chemical fertilizers. Biofertilizers are superior alternatives to synthetic fertilizers, as they enhance soil quality and contain minerals that are essential for plant fertility and productivity.

## 5.13 Conclusions and Future Perspectives

Nanofertilizers have garnered significant attention due to their unique characteristics and practical applications in agricultural systems. From the explanation above, it is clear that while slow-release fertilizers are more effective at supplying a range of nutrients in appropriate amounts, nanoparticle fertilizers also enhance nutrient usage efficiency by avoiding nutrient fixation or loss to the environment. Additionally, they shield the crops from bacterial and fungal infections, but they could unintentionally harm nontarget plant-associated microorganisms that are crucial to plant nutrition, including mycorrhizal communities or bacteria that fix nitrogen. Furthermore, this broad-spectrum antibacterial treatment may have an adverse effect on beneficial soil microbes that break down organic matter and preserve soil fertility over the long run. If that is the case, regular exposure to these nanoformulation treatments may have unfavorable long-term effects on plant yields and soil fertility. It is possible to optimize the size and concentrations of NP to get the desired results. One of the primary causes of concern is the assimilation of NP and its consequent accumulation in the food web. Consequently, it is imperative to pursue the development of experimental models that elucidate the interactions between plants and animals, as well as the impacts that are investigated at the individual atrophic level. Plant-NP interactions modify plant cell gene and protein profiles, which impact biological processes and eventually change how plants grow and develop. Therefore, studies must be conducted to gather data regarding the molecules involved in the NP's absorption and translocation. While studies on metallic nanoparticles' distribution and application processes are still in their early stages, their potential benefits for sustainable agriculture have not been fully explored.

## References


Abdel-Aziz H, Hasaneen MN, Omar A (2018) Effect of foliar application of nano chitosan NPK fertilizer on the chemical composition of wheat grains. Egypt J Bot 58:87–95

Adisa IO, Pullagurala VLR, Peralta-Videa JR, Dimkpa CO, Elmer WH, Gardea-Torresdey JL, White JC (2019) Recent advances in nano-enabled fertilizers and pesticides: a critical review of mechanisms of action. Environ Sci Nano 6:2002–2030

Alharby HF, Metwali EM, Fuller M, Aldhebiani AY (2016) Impact of application of zinc oxide nanoparticles on callus induction, plant regeneration, element content and antioxidant enzyme activity in tomato (Solanum lycopersicum Mill.) under salt stress. Arch Biol Sci 68:723–735

Ali SM, Yousef NMH, Nafady NA (2015) Nafady "application of biosynthesized silver nanoparticles for the control of land snail *Eobaniavermiculata* and some plant pathogenic fungi". J Nanomater 2015:218904

Al-Juthery HW, Al-Fadhly JT, Ali EAHM, Al-Taee RAHG (2019) Role of some nanofertilizers and atonikin maximizing for production of hydroponically-grown barley fodder. Int J Agric Stat Sci 15:565–570

Bhardwaj AK, Arya G, Kumar R, Hamed L, Pirasteh-Anosheh H, Jasrotia P, Kashyap PL, Singh GP (2022) Switching to nanonutrients for sustaining agroecosystems and environment: the challenges and benefits in moving up from ionic to particle feeding. J Nanobiotechnol 20:19

Chhipa H, Joshi P (2016) Nano-fertilizers, nanopesticides and nanosensors in agriculture. In: Ranjan S, Dasgupta N, Lichtfouse E (eds) Nanoscience in food and agriculture, Sustainable agriculture reviews, vol 20, pp 247–282

Dikshit A, Shukla SK, Mishra RK (2013) Exploring nanomaterials with PGPR in current agricultural scenario. Lap Lambert Academic Publishing Company, Germany. Print

Dimkpa CO, Bindraban PS (2016) Fortification of micronutrients for efficient agronomic production: a review. Agron Sustain Dev 36(1):7

Dimkpa CO, White JC, Elmer WH, Gardea-Torresdey J (2017) Nanoparticle and ionic Zn promote nutrient loading of sorghum grain under low NPK fertilization. J Agric Food Chem 65:8552–8559

Drostkar E, Talebi R, Kanouni H (2016) Foliar application of Fe, Zn and NPK nano-fertilizers on seed yield and orphological traits in chickpea under rainfed condition. J Res Ecol 4:221–228

Duhan JS, Kumar R, Kumar N, Kaur P, Nehra K, Duhan S (2017) Nanotechnology: the new perspective in precision agriculture. Biotechnol Rep (Amst) 15:11–23

El-Saadony MT, El-Wafai NA, El-Fattah HIA, Mahgoub SA (2019) Biosynthesis, optimization and characterization of silver nanoparticles using a soil isolate of Bacillus pseudomycoides MT32 and their antifungal activity against some pathogenic fungi. Adv Anim Vet Sci 7:238–249

El-Saadony MT, El-Hack A, Mohamed E, Taha AE, Fouda MM, Ajarem JS, Maodaa NS, Allam AA, Elshaer N (2020) Ecofriendly synthesis and insecticidal application of copper nanoparticles against the storage pest Triboliumcastaneum. Nano 10:587

El-Saadony MT, Alkhatib FM, Alzahrani SO, Shafi ME, Abdel-Hamid SE, Taha FT, Aboelenin SM, Soliman MM, Ahmed NH (2021) Impact of mycogenic zinc nanoparticles on performance, behavior, immune response, and microbial load in Oreochromisniloticus. Saudi J Biol Sci 28:4592–4604

Fraceto LF, Grillo R, de Medeiros GA, Scognamiglio V, Rea G, Bartolucci C (2016) Nanotechnology in agriculture: which innovation potential does it have? Front Environ Sci 4:20
Godfray HCJ, Beddington JR, Crute IR, Haddad L, Lawrence D, Muir JF, Pretty J, Robinson S, Thomas SM, Toulmin C (2010) Food security: the challenge of feeding 9 billion people. Science 327:812–818
Guru T, Veronica N, Thatikunta R, Reddy SN (2015) Crop nutrition management with nano fertilizers. Int J Environ Sci Technol 1(1):4–6
Hoang SA, Nguyen LQ, Nguyen NH, Tran CQ, Nguyen DV, Le NT, Ha CV, Vu QN, Phan CM (2019) Metal nanoparticles as effective promotors for maize production. Sci Rep 9:13925
Jampílek J, Kralova K (2017) 2017 nanomaterials for delivery of nutrients and growth-promoting compounds to plants. In: Prasad R, Kumar M, Kumar V (eds) Nanotechnology: an agricultural paradigm. Springer, Singapore, pp 177–226
Jiang M, Song Y, Kanwar MK, Ahammed GJ, Shao S, Zhou J (2021) Phytonanotechnology applications in modern agriculture. J. Nanobiotechnol. 19:430
Josef J, Katarína K (2015) Application of nanotechnology in agriculture and food industry, its prospects and risks. Ecol Chem Eng 22(3):321–361
Kah M, Machinski P, Koerner P, Tiede K, Grillo R, Fraceto LF, Hofmann T (2014) Analysing the fate of nanopesticides in soil and the applicability of regulatory protocols using a polymer-based nanoformulation of atrazine. Environ Sci Pollut Res Int 21:11699–11707
Kaushik S, Djiwanti SR (2017) Nanotechnology for enhancing crop productivity. In: Prasad R, Kumar M, Kumar V (eds) Nanotechnology: an agricultural paradigm. Springer, Singapore, pp 249–262
Kole C et al (2013) Nanobiotechnology can boost crop production and quality: first evidence from increased plant biomass, fruit yield and phytomedicine content in bittermelon (Momordicacharantia). BMC Biotechnol 13:37
Komarneni S (2009) Potential of nanotechnology in environmental soil science. Proceeding of 9th International Conference East and Southeast Asia Federation of Soil Science Societies, Korean Society of Soil Science and Fertilizers, Seoul. p. 27–30
Kottegoda N, Sandaruwan C, Priyadarshana G, Siriwardhana A, Rathnayake UA, BerugodaArachchige DM, Kumarasinghe AR, Dahanayake D, Karunaratne V, Amaratunga GA (2017) Urea hydroxyapatite nanohybrids for slow release of nitrogen. ACSNano 11(2):1214–1221
Li HJ, Du JZ, Liu J et al (2016) Smart superstructures with ultrahigh pH-sensitivity for targeting acidic tumor microenvironment: instantaneous size switching and improved tumor penetration. ACS Nano 10:6753–6761
Liu R, Lal R (2014) Synthetic apatite nanoparticles as a phosphorus fertilizer for soybean (Glycine max). Sci Rep 4:5686
Liu XM, Zhang FD, Zhang SQ, He XS, Fang R, Feng Z, Wang YJ (2010) Effects of nano-ferric oxide on the growth and nutrients absorp-tion of peanut. Plant Nutr Fertilizing Sci 11:14–18
Liu R, Kang Y, Pei L, Wan S, Liu S, Liu S (2016) Use of a new controlled-loss-fertilizer to reduce nitrogen losses during winter wheat cultivation in the Danjiangkou reservoir area of China. Commun Soil Sci Plant Anal 47:1137–1147
López-Valdez F, Miranda-Arámbula M, Ríos-Cortés AM, Fernández-Luqueño F, de la Luz V (2018) Nanofertilizers and their controlled delivery of nutrients. In: López-Valdez F, Fernández-Luqueño F (eds) Agricultural nanobiotechnology. International Publishing AG, Cham, Switzerland, pp 35–48
Lu L, Huang M, Huang Y, Corvini PF-X, Ji R, Zhao L (2020) Mn3O4 nanozymes boost endogenous antioxidant metabolites in cucumber (Cucumis sativus) plant and enhance resistance to salinity stress. Environ Sci Nano 7:1692–1703
Macht F, Eusterhues K, Pronk GJ, Totsche KU (2011) Specific surface area of clay minerals: comparison between atomic force microscopy measurements and bulk-gas (N2) and liquid (EGME) adsorption methods. App Clay Sci 53(1):20–26
Mahapatra DM, Satapathy KC, Panda B (2022) Biofertilizers and nanofertilizers for sustainable agriculture: phycoprospects and challenges. Sci Total Environ 803:149990

Malusa E, Vassilev N (2014) A contribution to set a legal framework for biofertilisers. Appl Microbiol Biotechnol 98:6599–6607

Malusá E, Sas-Paszt L, Ciesielska J (2012) Technologies for beneficial microorganisms inocula used as biofertilizers. ScientificWorldJournal 2012:491206

Marschner H (ed) (2011) Marschner's mineral nutrition of higher plants. Academic press

Mishra S, Singh BR, Singh A, Keswani C, Naq-vi AH, Singh HB (2014) Bio-fabricated silver nanoparticles act as a strong fungicide against *Bipolarissorokiniana* causing spot blotch disease in wheat. PLoS One 9:e97881

Mittal D, Kaur G, Singh P, Yadav K, Ali SA (2020) Nanoparticle-based sustainable agriculture and food science: recent advances and future outlook. Front Nanotechnol 2:10

Monreal C, DeRosa M, Mallubhotla S, Bindraban P, Dimkpa C (2016) Nanotechnologies for increasing the crop use efficiency of fertilizer-micronutrients. Biol Fertiler Soils 52(3):423–437

Morales-Díaz AB, Ortega-Ortíz H, Juárez-Maldonado A, Cadenas-Pliego G, González-Morales S, Benavides-Mendoza A (2017) Application of nano-elements in plant nutrition and its impact in ecosystems. Adv Nat Sci Nanosci Nanotechnol 8:013001

Mushtaq W, Shakeel A, Fazili MA, Chakrabartty I, Sevindik M (2020) Pros and cons of nanotechnology. Nanobiotechnol Agric:207–222. https://doi.org/10.1007/978-3-030-39978-8_13

Nibin PM, Ushakumari K, Ishrath PK (2019) Organic nano NPK formulations on soil microbial and enzymatic activities on post harvest soil of Bhindi. Int J Curr Microbiol App Sc 8(04):1814–1819

Prasad A, Astete CE, Bodoki AE, Windham M, Bodoki E, Sabliov CM (2017) Zein nanoparticles uptake and translocation in hydroponically grown sugar cane plants. J Agric Food Chem 66:6544–6551

Preetha PS, Balakrishnan N (2017) A review of nano fertilizers and their use and functions in soil. Int J Curr Microbiol Appl Sci 6:3117–3133

Raddy R, Salimath M, Geetha K, Shankar A (2018) ZnO nanoparticle improves maize growth, yield and seed zinc under high soil pH condition. Int J Curr Microbiol Appl Sci 7:1593–1601

Rahmatpour S, Shirvani M, Mosaddeghi MR et al (2017) Dose–response effects of silver nanoparticles and silver nitrate on microbial and enzyme activities in calcareous soils. Geoderma 285:313–322

Rajaie M, Ziaeyan AH (2009) Combined effect of zinc and boron on yield and nutrients accumulation in corn. Int J Plant Product 3:35–440

Rajiv P, Vanathi P (2018) Effect of parthenium based vermicompost and zinc oxide nanoparticles on growth and yield of Arachishypogaea L. in zinc deficient soil. Biocatal Agric Biotechnol 13:251–257

Rajput VD, Minkina T, Sushkova S, Tsitsuash-vili V, Mandzhieva S, Gorovtsov A, Nevidomskyaya D, Gromakova N (2018) Effect of nano particles on crops and soil microbial communities. Soils Sediments 18:2179–2187

Raliya R, Saharan V, Dimkpa C, Biswas P (2018a) Nanofertilizer for precision and sustainable agriculture: current state and future perspectives. J Agric Food Chem 66:6487–6503

Raliya R, Saharan V, Dimpka C, Biswas P (2018b) Nanofertilizer for precision and sustainable agriculture: current state and future perspectives. J Agric Food Chem 66(26):6487–6503

Rawat M, Nayan R, Negi B, Zaidi M, Arora S (2017) Physio-biochemical basis of iron-sulfide nanoparticle induced growth and seed yield enhancement in B. juncea. Plant Physiol Biochem 118:274–284

Reda FM, El-Saadony MT, Elnesr SS, Alagawany M, Tufarelli V (2020) Effect of dietary supplementation of biological curcumin nanoparticles on growth and carcass traits, antioxidant status, immunity and caecal microbiota of Japanese quails. Animals 10:754

Reda FM, El-Saadony MT, El-Rayes TK, Attia AI, El-Sayed SA, Ahmed SY, Madkour M, Alagawany M (2021) Use of biological nano zinc as a feed additive in quail nutrition: biosynthesis, antimicrobial activity and its effect on growth, feed utilisation, blood metabolites and intestinal microbiota. Ital J Anim Sci 20:324–335

Saharan V, Kumaraswamy RV, Choudhary RC, Kumari S, Pal A, Raliya R, Biswas P (2016) Cu-chitosan nanoparticle mediated sustainable approach to enhance seedling growth in maize by mobilizing reserved food. J Agric Food Chem 64:6148–6155

Schlich K, Hund-Rinke K (2015) Influence of soil properties on the effect of silver nanomaterials on microbial activity in five soils. Environ Pollut 196:321–330

Seleiman MF, Almutairi KF, Alotaibi M, Shami A, Alhammad BA, Battaglia ML (2020) Nano-fertilization as an emerging fertilization technique: why can modern agriculture benefit from its use? Plan Theory 10(1):2

Shankramma K, Yallappa S, Shivanna MB, Manjanna J (2016) Fe2O3 magnetic nanoparticles to enhance S. lycopersicum (tomato) plant growth and their biomineralization. Appl Nanosci 6:983–990

Shukla SK et al (2015) Prediction and validation of gold nanoparticles (GNPs) on plant growth promoting rhizobacteria (PGPR): a step toward development of nanobiofertilizers. Nanotechnol Rev 4:439–448

Shukla P, Chaurasia P, Younis K, Qadri OS, Faridi SA, Srivastava G (2019) Nanotechnology in sustainable agriculture: studies from seed priming to post-harvest management. Nanotechnol Environ Eng 4:11

Simarmata T, Hersanti, Turmuktini T, Fitriatin BN, Setiawati M, Purwanto (2016) Application of bioameliorant and biofertilizers to increase the soil health and rice productivity. Hayati J Biosci 23:181–184

Singh BK, Sarma C, Keswani (eds) (2016) Agriculturally important microorganisms: commercialization and regulatory requirements in Asia. Springer, Singapore, pp 133–145

Singh MD, Chirag G, Prakash PO, Moh-an MH, Prakasha G, Vishwajith. (2017) Nano-fertilizers is a new way to increase nutrients use efficiency in crop production. Int J Agric Sci 9(7):3831–3833

Subbaiah LV, Prasad TNVKV, Krishna TG, Sudhakar P, Reddy BR, Pradeep T (2016) Novel effects of nanoparticulate delivery of zinc on growth, productivity, and zinc biofortification in maize (Zea mays L.). J Agric Food Chem 64:3778–3788

Subramanian KS, Tarafdar JC (2009) Nano technology in soil science. Proceedings of the Platinum Jubilee Symposium on Soil Science. IARI Campus, New Delhi, pp 326–334

Tarafdar JC, Raliya R, Tathore I (2012) Microbial synthesis of phosphorous nanoparticle from tri-calcium phosphate using *Aspergillus tubingensis*TFR-5. J Bionanosci 6:84–89

VandeVoort AR, Arai Y (2019) The role of nanotechnology in the fortification of plant nutrients and improvement of crop production. Appl Sci 9:499

Xiong L, Wang P, Hunter MN, Kopittke PM (2018) Bioavailability and movement of hydroxyapatite nanoparticles (HA-NPs) applied as a phosphorus fertiliser in soils. Environ Sci Nano 5:2888–2898

Xuan L, Zhao J, Skonieczna M, Zhou PK, Huang R (2023) Nanoparticles induced potential toxicity on human health: application, toxicity mechanisms, and evaluation models. MedComm 4(4):e327

Yuan J, Chen Y, Li H, Lu J, Zhao H, Liu M, Nechitaylo GS, Glushchenko NN (2018) New insights into the cellular responses to iron nanoparticles in Capsicum annuum. Sci Rep 8:3228

# Chapter 6
# Opportunities and Challenges of Nanofertilizer Use in Agriculture

**Manoj Kumar, Tejendra Krishna, Satdev, Nintu Mandal, and Suparna Biswas**

**Abstract** A nanometer is a unit of measurement equal to one-billionth of a meter. The range of measurements from 1 to 100 nanometers is commonly referred to as the 'nanoscale.' Nanotechnology has shown remarkable potential across various fields like medicine, pharmaceuticals, energy, materials science, and plant science, leading to its increased popularity. With the escalating global demand for food, the widespread use of chemical fertilizers has resulted in detrimental effects on soil, the environment, and human health. As an effective alternative, nanofertilizers have emerged, comprising carefully regulated nanoparticles delivering macro- and micronutrients to the plant rhizosphere. These synthetic substances incorporate essential nutrients like nitrogen, phosphorus, potassium, iron, and manganese, either individually or combined with nano-sized adsorbents. Due to their capability to enhance nutrient uptake efficiency, nanofertilizers offer superior nutritional management. Current agricultural practices, primarily established during the green revolution, have become unsustainable, especially given population growth and the impacts of climate change. To ensure food security, sustainable agriculture offers a more resilient, efficient, and enduring system. Nanotechnology will significantly contribute to advancing this transition. In order to improve crop nutrient efficiency, we've compiled a list of promising opportunities and methodologies for applying nanotechnology. Adopting a systems approach is crucial in designing potential nanotechnologies, considering possible obstacles in their development and implementation.

M. Kumar (✉)
ICAR-National Research Centre for Makhana, Darbhanga, Bihar, India

T. Krishna · Satdev · N. Mandal
Department of Soil Science and Agricultural Chemistry, Bihar Agricultural University, Bhagalpur, Bihar, India

S. Biswas
Department of Physics, Indian Institute of Engineering Science and Technology, Howrah, West Bengal, India

 
A. K. Patra (ed.), *Nanofertilizers in Agriculture*, Sustainability Sciences in Asia and Africa, https://doi.org/10.1007/978-981-97-8112-6_6

**Keywords** Nanotechnology · Nanoscale · Climate change · Nanofertilizer · Nutrient use efficiency · Sustainable agriculture

## 6.1 Introduction

Fertilizers, often referred to as plant nutrients, are substances that provide essential elements or minerals necessary for plant growth and development. In agriculture, the primary focus of nanotechnology lies in the development of ecologically sound nanofertilizers (NFs) as an alternative to traditional fertilizers. The excessive use of fertilizers has depleted soil fertility by eradicating beneficial bacteria. Efforts are underway to create nanodimensional nutrient carriers or transporters ranging from 1 to 100 nanometers (Abobatta 2018). NFs can be generated from bulk fertilizer materials or by encapsulating other elements in nanomaterials (NMs). Nanoparticles possess a large surface area and the ability to retain and gradually release nutrients, allowing plants to absorb nutrients as needed without the negative side effects associated with traditional fertilizers (Preetha and Balakrishnan 2017). This additional surface area aids in various metabolic processes, enhancing photosynthesis and ultimately increasing productivity. NFs not only reduce ecotoxicity and prevent nutrient loss to soil or groundwater but also encourage better absorption and utilization efficiency (Tarafdar and Raliya 2013). The smaller size of nanoparticles allows for enhanced plant penetration, facilitating improved nutrient uptake. Consequently, NFs significantly enhance photosynthesis, nutrient uptake, translocation, and pest resistance, leading to improved agricultural yields and soil quality. Nano-fertilizers offer a more efficient and environmentally friendly alternative to chemical fertilizers, owing to their physiochemical characteristics (Zulfiqar et al. 2019). Nanoparticles (NPs) play a crucial role in sustainable agriculture due to their smaller size and larger surface area, enabling the development of nanopesticides, nanoinsecticides, and nanofertilizers (Raguraj et al. 2020). Recent studies have highlighted the potential use of nanoparticles in managing plant pests like weeds, fungi, and insects (Sood 2021). The application of NFs is projected to significantly boost crop output by improving nutrient accessibility for plants (Abdelmigid et al. 2022). Nanotechnology shows promise in creating nanofertilizers with greater effectiveness, minimal ecological risks, and lower costs compared to their inorganic counterparts. These nanofertilizers, sourced from various elements or compounds, offer reduced nutrient losses, improved plant uptake, and biodegradability, thus differing from conventional fertilizers (León-Silva et al. 2018; Solanki et al. 2015). Conventional fertilizers pose concerns regarding massive nutrient losses from leaching and volatilization, issues that nanofertilizers aim to mitigate (Huang et al. 2017; Pan et al. 2016). Despite their potential to enhance soil fertility, plant nutrition, distribution, and protection, the widespread adoption of nanofertilizers in agriculture faces several hurdles. First, developing nano-agrochemicals that promote environmental sustainability is challenging due to the nascent nature of nanotechnology. Second, barriers related to pricing, regulation, and marketing hinder the potential applications of nanoagrochemicals. Third, questions persist regarding their usage in agriculture, mainly concerning eco-toxicity, long-term impacts, and environmental consequences due to insufficient research (Dubey and Mailapalli 2016). This chapter delves into the use of nanofertilizers in agricultural fields. Nanofertilizers can be tailored to meet the specific nutrient requirements of plants, drawing from a diverse range of elements and compounds. Previous research indicates the beneficial effects of various nanoparticles: silver nanoparticles enhanced the germination of *Boswellia ovalifoliolata* (Savithramma et al. 2012), ZnO nanoparticles notably improved the growth characteristics and biochemistry of cotton (Venkatachalam et al. 2017), CuO-NPs stimulated rice growth (Da-Costa and Sharma 2016), and Cu nanoparticles had a positive impact on wheat (Hafeez et al. 2015). Studies have also shown the efficacy of ZnO nanoparticles on maize (Taheri et al. 2016), Cu and Zn nanoparticles on wheat (Taran et al. 2017), and silica nanoparticles on basil (Kalteh et al. 2018). Significant effect on crop production and slow released pattern DTPA Zn were recorded with zinc containing nanoclay polymers composite (ZnNCPCs) over commercial Zn fertilizer (Adhikary et al. 2020a, b; Dubey et al. 2021; Mandal et al. 2015; Mandal et al. 2019; Mandal et al. 2021a, b). Rokana et al. (2023) conducted a comparative study revealing that the release pattern of DTPA Fe from the iron-containing nano clay polymer composite (Fe-NCPC) surpassed that of commercial $FeSO_4.7H_2O$. Mandal et al. (2018a, b) also reported that novel chitosan grafted Zn containing nanoclay polymer bio-composite (CZNCPBC) proved to be promising as an innovative controlled release $Zn^{2+}$ fertilizer. Despite their considerable benefits in enhancing soil fertility, managing plant nutrition, facilitating smart distribution, and providing protective qualities, nanofertilizers have not seen widespread adoption as an alternative to conventional counterparts in agriculture. This study aims to clarify the usage of nanofertilizers in mitigating losses and elaborates on the various studies demonstrating the potential impacts of different nanoparticles on plant growth and development.

## 6.2 Nanoformulation

Chemical fertilizers are commonly applied to plants to boost growth and productivity. These fertilizers typically provide N, P, and K as the major essential nutrients for various crops and diverse growing conditions. Nitrogen supports leaf development, stimulating protein and chlorophyll production; phosphorus aids in root, flower, and fruit growth; and potassium contributes to stem, root, and protein synthesis (Mandal et al. 2009). Nanofertilizers, a type of substance capable of storing a significant amount of nutrients for gradual release, have emerged. They facilitate easier absorption of necessary nutrients by crops without causing harm from specially formulated fertilizer inputs (Ghormade et al. 2011). At the nanoscale (1–100 nm), materials exhibit distinct physical, chemical, and biological characteristics, setting them apart from larger-scale particles. In agriculture, one primary application of nanotechnology lies in the distribution of agricultural chemicals, particularly fertilizers (Manimegalai et al. 2011). Naturally occurring nanoparticles like zeolite minerals have found use in agriculture. However, engineered nanomaterials are synthesized to meet specific chemical and physical property requirements for various applications. Nanofertilizers are under scrutiny for their potential to enhance plant nutrition and nutrient efficiency compared to conventional fertilizers, as supported by research publications. Different nanoformulations are illustrated in figure 6.1. Three main categories of nanofertilizers have been proposed: (1) nanoscale coatings (applying nanoparticles to conventional fertilizers); (2) nanoscale additives (nanoparticles containing nutrients); and (3) nanoscale fertilizers (nanoparticles holding nutrients) (Mastronardi et al. 2015).

**Nanoscale fertilizer:** The fertilizer is applied in the form of nanoscale particles or emulsions because reducing the input size improves uptake and overall release efficiency, providing the necessary nutrition for plants with potentially lower toxicity. This group comprises particles made of urea, ammonium salts, peat, and other conventional fertilizers.

**Nanoscale additives:** These materials replace a small quantity of macro scale inputs to enhance their qualities rather than directly substituting the nutrient itself. The nanomaterials may function as micronutrients, fertilizers, or adjuvant materials that act as binders or water-retentive substances in relation to plant growth. The primary purposes of adding nanoscale fertilizer additions to soil are to confer antimicrobial or insect resistance to plants.

**Nanoscale coating:** In this field, a film or nanoscale coating encapsulates the macroscale fertilizer. The presence of nanoscale pores or gaps in the film facilitates the delayed release of nutrient solubility. While still in their infancy in agriculture, these applications are already widely used in engineering and medicine. Another remarkable application of nanotechnology is the encapsulation of beneficial microorganisms, which effectively enhance plant root growth and serve health-related objectives. Various bacteria or fungi used in this application can improve the availability of potassium, nitrogen, and phosphorus in the plant's root zone. Clays such as kaolinites, smectites, halloysites, and palygorskites are utilized in fertilizer products.

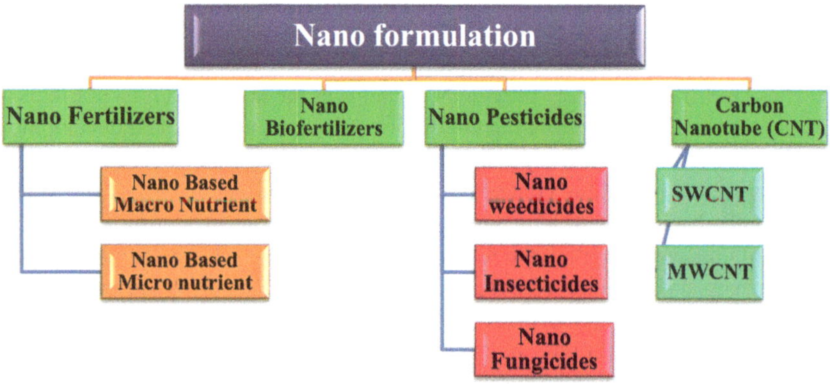


**Fig. 6.1** Different nanoformulations

## 6.3 Nanoscale Fertilizer

### *6.3.1 Nano-Based Nitrogen Fertilizer (N-NF)*

Nitrogen stands out as the pivotal nutrient crucial for plant growth. Conventional fertilizers manage only 50–70% of the necessary nitrogen, and soil leaching contributes to nitrogen absorption, reducing its effectiveness (referred to as NUE) for plants. Previous attempts to improve NUE in standard fertilizers fell short, even with slow-release polymer-coated nitrogen fertilizers, which weren't as potent as nanotechnology-based ones (Hossain et al. 2008a). Studies incorporated urease enzymes into nanoporous silica, controlling nitrogen release via urea hydrolysis (Hossain et al. 2008b). Losses in the environment occur when crops can't utilize nitrogen, leading to significant resource, financial, and environmental losses. To address this, certain fertilizer properties need consideration: adequate application, deep placement, granular urea, understanding crop response, and employing slow-release nanofertilizers. Urea, a nitrogen-rich source, is pivotal, favoring gradual nitrogen delivery to plants. Innovative fertilizers like Urea-Hydroxyapatite Nanohybrids interact urea with HA-NPs, releasing crucial nitrogen gradually. This slow-release quality was suggested to enhance rice crop output at 50% lower urea concentrations (Kottegoda et al. 2014, 2016, 2017). Urea-coated nanohydroxyapatite composites, researched for seed coatings, showed improved seasonal crop germination (Pabodha et al. 2022). Urea-silica nanohybrids offer precise silica and nitrogen delivery, exhibiting larger nitrogen loading than other nitrogen-releasing fertilizers (Siriwardena et al. 2017; De-Silva et al. 2020). Zeolite-based nitrogen nanofertilizers enhance nitrogen uptake, improving plant growth and soil properties compared to traditional fertilizers (Rajonee et al. 2016). Studies evaluating K Nano fertilizers and N biofertilizers on red bean production demonstrated their positive influence on yield and yield components. Combining N biofertilizer and Khazraa K chelate Nanofertilizer yielded higher red bean yields, showcasing their potential as alternatives to chemical fertilizers, being more economical and environmentally friendly (Farnia and Ghorbani 2014). Urea-hydroxyapatite nanohybrids significantly improved soil P, leaf N, and P content in Low Country tea plants, boosting *Camellia sinensis* (L.) Kuntze tea yields, potentially reducing excessive urea usage (Raguraj et al. 2020). Nano calcium carbonate-based slow-release fertilizers showed five times lesser urea solubility compared to pure urea, indicating controlled release properties beneficial for agriculture (Rathnaweera et al. 2019).

### *6.3.2 Nano-Based Phosphorus Fertilizer (P-NF)*

The second crucial nutrient for plants is phosphorus (P). Commercially available water-soluble phosphate salts like TSP (triple superphosphate), DAP (diammonium phosphate), and MAP (monoammonium phosphate) play a significant role in providing this nutrient. However, the influx of phosphorus into water systems poses a considerable challenge, prompting the proposal of remediation technologies aimed at reducing fertilizer application and preventing phosphorus from leaching into water bodies (De-Bashan and Bashan 2004). Nanostructuring phosphorus through surface fictionalization using nanotechnology alters its chemical properties, diminishing its soil mobility and the bioavailability of algae. Developing phosphorus-based nanofertilizers alongside conventional ones could enhance surface water quality, elevate agricultural production, and improve phosphorus use efficiency (Liu and Lal 2014). Research has shown that treating soybean plants with nanosized hydroxyapatite, a form of phosphorus, instead of conventional P fertilizers like $Ca(H_2PO_4)_2$, resulted in 30% taller plants. Phosphorus is pivotal for proper growth in soybeans and other legumes, regulating crucial plant functions. Studies by Samavini et al. (2018) have focused on creating an effective phosphorus nutrient system to address inherent issues with existing phosphorus fertilizers. Strategies like wet chemical precipitation have been employed to produce hydroxyapatite nanoparticles surface-modified by siliceous acid. These studies used *Zea mays* as a model crop to examine water release and bioavailability, revealing increased phosphorus availability in the presence of organic acids compared to pure hydroxyapatite nanoparticles or rock phosphate. Moreover, integrating nitrogen (N), phosphorus (P), and potassium (K) into chitosan nanoparticles to form NPK nanofertilizer has shown promising results in enhancing nutrient uptake, photosynthesis, and coffee plant growth (Ha et al. 2019). Application of this nanofertilizer resulted in significantly higher potassium (67.50%), nitrogen (17.04%), and phosphorus (16.31%) content in treated leaves compared to controls. Additionally, coffee seedlings exhibited increased total chlorophyll content, higher photosynthetic rates, more leaves, greater plant heights, and larger leaf areas. These findings suggest that nanofertilizers hold potential for enhancing the efficacy of coffee fertilization. Using chitosan nanoparticles created by ionic gelation with TPP to produce NPK nanofertilizer could offer a regulated slow-release carrier system. These developments indicate the potential of NPK nanofertilizers in fostering environmentally friendly and sustainable agriculture in the near future.

### 6.3.3 Nano-Based Potassium Fertilizer (K-NF)

Potassium nanofertilizers (K-NFs) exert significant control over crucial plant functions such as water flow, photosynthetic capacity augmentation, cell tissue fortification, nitrate absorption, flowering stimulation, and synthesis of carbohydrates and enzymes within plant systems. Recent global reports have highlighted the superior crop response of K-NFs compared to conventional potassium fertilizers (Hussain et al. 2019; Nido et al. 2019). Ghahremani et al. (2014) demonstrated that applying nano-K at a concentration of 1/1000 was highly effective in increasing basil crop yield, leaf area, and chlorophyll content. Similarly, summer squash treated with foliar nano-K fertilizers displayed enhanced product quality, insect resistance, drought tolerance, and increased leaf quantity (Gerdini 2016). Studies have shown that K-NF lithovit boosts photosynthesis, thereby promoting plant growth and improving yield (Attia et al. 2016). Furthermore, the utilization of chitosan and methacrylic acid NFs instead of traditional K-fertilizers stimulated root elongation in garden peas (Khalifa and Hasaneen 2018). Similarly, the application of K-NFs at 150 ppm enhanced nutrient uptake in peanut plants (Afify et al. 2019).

### *6.3.4 Nano-Based Calcium Fertilizer (Ca-NFs)*

Calcium nanofertilizers (Ca-NFs) play pivotal roles in stabilizing cell walls, retaining minerals in soil and during transit, neutralizing hazardous chemicals, and promoting seed growth, all hinging on the presence of calcium (Ca). When compared to higher concentrations of calcium chloride ($CaCl_2$), even a small quantity of Ca-NF has been shown to boost both yield and quality of pomegranates (Davarpanah et al. 2018). Similarly, applying Ca-NFs during the pre-harvest stage of apple fruit has been found to significantly enhance both quantitative and qualitative traits compared to $CaCl_2$ (Ranjbar et al. 2018). Moreover, in studies involving a 500 mg per liter spray of Ca-NFs, a 15-day advancement in flowering over the control treatment resulted in a substantial 56.3% increase in flowering (Seydmohammadi et al. 2020). In the cultivation of groundnut crops, the application of $CaCO_3$-NFs (ranging from 20 to 80 nanometers, at a concentration of 160 mg $L^{-1}$ of Ca) was compared to $Ca(NO_3)_2$ (200 mg $L^{-1}$). This comparison revealed a significant increase in groundnut fresh biomass, calcium absorption, and transfer from root to shoot upon the utilization of Ca-NFs (Liu and Lal 2015). Moreover, in managing nitrogen consumption in grapevines, the inclusion of urea with calcium phosphate nanoparticles ($Ca_3(PO_4)_2.nH_2O$) as a nitrogen source has been explored (Gaiotti et al. 2021).

### *6.3.5 Nano-Based Magnesium Fertilizer (Mg-NF)*

Magnesium (Mg) serves as the foundational element in chlorophyll, essential for photosynthesis and crucial for overall plant development (Kumar 2022). Due to its high mobility and tendency to leach from soil, magnesium is frequently depleted. Shinde et al. (2018) found that the use of magnesium hydroxide NPs at a concentration of 500 mg $kg^{-1}$ resulted in 100% seed germination and notably improved maize plant growth. Additionally, in a field experiment, foliar application of Mg-NPs and Fe-NPs at 0.5 g $L^{-1}$ enhanced photosynthetic efficiency, leading to increased cowpea growth and productivity (Delfani et al. 2014).

### *6.3.6 Nano-Based Sulfur Fertilizer (S-NF)*

Sulfur nanofertilizers (S-NFs) are instrumental in enhancing plant defenses, optimizing nitrogen utilization efficiency, and aiding chlorophyll production. While elemental sulfur (S) contains sulfur, it's typically not readily available for agricultural plants until undergoing biological oxidation. Valle et al. (2019) highlighted that the size of fertilizer particles largely influences this oxidation process. Excessive application of sulfur leads to increased leakage into the soil, posing detrimental effects on the ecosystem. The advent of S-NFs presents a viable alternative to conventional sulfur fertilizers. Nano-coating S (<100 nm layer) onto nitrogen and phosphorus fertilizers facilitates their gradual release, particularly beneficial in sulfur-deficient soils (Brady and Weil 1999). The utilization of biosynthesized S-NFs or *Ocimum basilicum* leaf extract on sunflower seedlings has demonstrated enhanced sulfur metabolism, increased seedling water content, and reduced physiological dehydration (Ragab and Saad-Allah 2020). Thirunavukkarasu et al. (2018) found that applying nano-sulfur at 30 kg $ha^{-1}$ effectively improves sulfur utilization efficiency, reducing the need for sulfur fertilizer by 25% while concurrently augmenting soil sulfur reserves. Moreover, in maize crops, supplementing NPK fertilizer with sulfur nutrients, converted into nanoparticles, can influence reproductive development (Dhlamini et al. 2020). Comparative evaluations by Selva et al. (2014) revealed that S-NF formulations release nutrients at a slower rate (1000–1200 h) compared to standard sulfur fertilizers (300–350 h).

### *6.3.7 Nano-Based Iron Fertilizer (Fe-NF)*

Fe nanofertilizers (Fe-NFs) provide a crucial source of iron, an essential cofactor for plant enzymes involved in diverse biological functions. Due to the insoluble nature of iron in soil, plants face limited accessibility to this nutrient. Utilizing remarkably stable slow-release NFs presents a promising strategy to make iron more available to plants. Notably, Fe-NFs are devoid of ethylene-based compounds that accelerate plant aging (Armin et al. 2014). These iron chelated NFs exhibit high stability and ensure a consistent iron release across a wide pH spectrum. Studies have demonstrated that the application of hematite NFs ($\alpha$-$Fe_2O_3$-NFs) and ferrihydrite NFs ($5Fe_2O_3.9H_2O$) to hydroponically grown maize and soybeans enhances chlorophyll levels in seedlings (Ghafariyan et al. 2013; Pariona et al. 2017). Delfani et al. (2014) found that foliar application of Fe-NFs (500 mg $L^{-1}$) increased cowpea leaf iron content by 34%, pod numbers by 47%, and chlorophyll content by 10% compared to non-nano Fe treatments. *Catharanthus roseus* treated with 30 μM-sized iron oxide NFs ($Fe_2O_3$) exhibited notable improvements in growth parameters, photosynthetic pigments, and total protein content (Askary et al. 2017). Hu et al. (2018) reported enhanced chlorophyll levels in watermelon and maize with $Fe_2O_3$ NPs (20–100 mg $L^{-1}$). Lower concentrations of ferrous oxide-NFs (0–0.75 g $L^{-1}$) applied to soybeans resulted in increased protein, lipid content, and chlorophyll (Sheykhbaglou et al. 2018). Application of root-applied $Fe_2O_3$-NFs (2–50 mg $L^{-1}$) boosted watermelon germination, development, and antioxidative enzymatic activities (Li et al. 2013). $Fe_2O_3$-NPs have also been shown to elevate auxin and malondialdehyde levels in rice and groundnut (Gui et al. 2015; Rui et al. 2016). *Medicago sativa* seedlings treated with nanoscale zero-valent iron (NZVI) exhibited increased chlorophyll levels (Kim et al. 2019). Additionally, the combination of $Fe_2O_3$ NFs with cornelian cherry fruit extract stimulated biomass in barley roots and shoots (Rostamizadeh et al. 2020). Glycoprotein and protein content in groundnut leaves saw an increase with seed treatment and soil application of $Fe_3O_4$-NFs (4000 mg $L^{-1}$) (Suresh et al. 2016). Rokana et al. (2022) conducted an assessment comparing the controlled release characteristics between chemically and biologically synthesized nano Fe over $FeSO_4.7H_2O$. Their findings indicated that biologically synthesized nano Fe exhibited a superior controlled release of DTPA Fe followed by chemical synthesis nano Fe in *Typic Haplustepts* and *Typic Ustifluvents* soils. Satdev et al. (2022) showed that the application of Fe-NCPC (Fe loaded nanoclay polymers composite) and chemical and biological synthesis nano Fe significantly improved the yield and nutrient composition in rice plants under Typic Haplustats.

### 6.3.8 Nano-Based Zinc Fertilizer (Zn-NF)

Zinc plays a crucial role as enzymes within plant systems rely on zinc for their catalytic activity (Kumar et al. 2016). Soil clay complexes often lead to poor zinc efficiency due to zinc adsorption, hindering its availability to plants. Wang et al. (2016) suggest that Zn-NFs effectively reduce zinc fixation in soil, facilitating targeted delivery to plants. Studies by Wang et al. (2013) highlight the positive impact of ZnO-NPs on specific crops compared to traditional zinc supplements, indicating their potential adoption by crop plants. Venkatachalam et al. (2017) and Faizan et al. (2018) observed that ZnO-NFs enhance antioxidant levels in tomatoes and aid cotton development. Narendhran et al. (2016) propose seed priming with Zn-NFs as an efficient and cost-effective method to supply zinc to crops. Zinc and zinc oxide (ZnO) are widely used in engineered nanomaterials for agriculture. Sooyeon et al. (2013) found that ZnO-NFs accelerated root growth in common buckwheat. Moghaddasi et al. (2017) noted an increase in cucumber plants' zinc content with soil application of ZnO-NFs (100 mg $kg^{-1}$). Seydmohammadi et al. (2020) demonstrated that topical application of ZnO-NPs on Lisianthus increased leaf chlorophyll, petal anthocyanin, and flower count. In mungbean, Mahajan et al. (2011) reported a significant boost in biomass production, root and shoot length by 42%, 41%, and 98%, respectively, with ZnO-NFs applied at 20 mg $L^{-1}$. Similarly, Zhao et al. (2014) found significant enhancements in cucumber growth and productivity with ZnO-NFs applied at 400 and 800 mg $kg^{-1}$. Lin and Xing (2007) observed increased root elongation in mustard and radish when treated with ZnO-NFs at 2 mg $L^{-1}$ compared to the control. Studies by Prasad et al. (2012); Sedghi et al.

(2013); Ramesh et al. (2014); Tarafdar et al. (2014) indicated that lower concentrations of ZnO-NFs enhanced seed germination in groundnut, soybean-grain legume, wheat-staple food crop, and bajra-small millet, respectively. Zhao et al. (2014) found that applying ZnO-NFs at 400 mg $kg^{-1}$ improved fruit quality, starch, globulin content, magnesium, and manganese levels in cucumber fruits compared to control treatments.

### 6.3.9 Nano-Based Copper Fertilizer (Cu-NF)

Cu plays a vital role in various physiological plant processes, including protein trafficking, cellular transport, antioxidative activity, and hormone signaling (Kumar 2022). Rawat et al. (2018) emphasize the significance of Cu-NFs as a primary source of copper for plants. Even at very low concentrations, Cu ions have been shown to enhance plant growth (Rajput et al. 2018). In onion cells undergoing active proliferation, Nagaonkar et al. (2015) observed an increase in the mitotic index through the application of biosynthesized citron fruit extract as Cu-NF (20 mg $L^{-1}$). Pradhan et al. (2015) demonstrated significant improvements in photosynthetic parameters and nitrogen and carbon assimilation in mungbean by employing CuO-NFs (0.05–1 mg $L^{-1}$) encapsulated in polyethylene glycol. Saharan et al. (2016) found that Cu-chitosan NF seed priming enhanced starch and protein mobilization and synthesis in maize seedlings. Sathiyabama and Manikandan (2018) reported increased growth and yield in finger millet with the application of Cu-chitosan NFs, either as a foliar spray or in seed coating. Similarly, Juarez-Maldonado et al. (2016) observed improved nutritional composition, increased lycopene pigment concentration in fruits, and enhanced growth and production in tomatoes through the use of chitosan hydrogel adjusted Cu-NFs. Yasmeen et al. (2017) noted elevated protein levels in wheat seeds treated with Cu-NFs, enhancing the grain's ability to withstand stress. Shende et al. (2017) demonstrated significant increases in root length and weight in pigeon pea seedlings treated with biogenic Cu NFs (20 nm size). Additionally, Shah and Belozerova (2009) found that applying Cu-NFs at doses of 130 and 600 mg $kg^{-1}$ led to a 40% and 91% growth enhancement in lettuce seedlings, respectively.

### 6.3.10 Nano-Based Manganese Fertilizer (Mn-NF)

Manganese (Mn) plays a critical role in both the anabolic processes of photosynthesis and the catabolic processes of respiration, along with nitrogen metabolism in plants. Research by Elmer and White (2016) highlights the superiority of Mn nanofertilizers (Mn-NFs) as a micronutrient source over commercially available $MnSO_4$ salt. These Mn-NFs have shown significant potential in enhancing mungbean growth and photosynthesis, surpassing control levels in shoot and root growth by 38% and 52%, respectively (Pradhan et al. 2013). Dimpka et al. (2018) demonstrated a 22% increase in manganese translocation efficiency in wheat crops with topical application of Mn-NFs. Similarly, Elmer and White (2016) reported a substantial 22% improvement in brinjal's economic output when treated with Mn-NFs compared to the control group.

### *6.3.11 Nano-Based Molybdenum Fertilizer (Mo-NF)*

Two enzymes essential for converting nitrate into nitrite and subsequently into ammonia require molybdenum (Mendel and Hansch 2002). Symbiotic nitrogen-fixing bacteria in legumes rely on molybdenum for the repair of atmospheric nitrogen (Awomi et al. 2012). Interest in examining the fertilizer potential of Mo-NPs has grown due to the appeal of nanofertilizers. An exploration into the effects of Mo-NPs as a fertilizer revealed intriguing outcomes. Tarafdar et al. (2012) utilized *Aspergillus tubingensis* TFR 29 to synthesize Mo-NPs (2–7 nm) at 4.0 ppm, resulting in a significant increase in microbial activity, beneficial enzymes, and root growth in the rhizosphere. In a pot experiment combining various Mo-NFs and nitrogen-fixing bacteria, Taran et al. (2014) discovered that applying these microorganisms and Mo-NFs concurrently notably amplified the microbial population in the rhizosphere.

### *6.3.12 Nano-Based Nickel and Silica Fertilizer (Ni-NF and Si-NF)*

Nickel plays a crucial role in numerous enzyme activities, maintaining cellular redox conditions, and governing various developmental processes, including growth responses, physiological changes, and biochemical reactions (Yusuf et al. 2011). In wheat crops, the application of Ni-NFs (5 nm) at a rate of 0.01 mg per liter increased the levels of chlorophyll 'a' and 'b' (Zotikova et al. 2018). Additionally, the utilization of $SiO_2$-NFs in rhizospheric soils has been found to enhance the soil microbial population and bolster productivity, protein synthesis, leaf area, and photosynthetic rates in maize (Suriyaprabha et al. 2012; Rangaraj et al. 2014).

## 6.4 Nanobiofertilizers

Biofertilizer preparations incorporate live formulations of beneficial microorganisms such as phosphate-solubilizing bacteria (PSB), blue-green algae (BGA), Rhizobium, mycorrhizal fungi, Azotobacter, and Azospirillum. Nanobiofertilizer formulations involve coating biofertilizers with nanoscale polymers, a process known as nanoencapsulation (Golbashy et al. 2017). As noted by Morales-Diaz et al. (2017), the bioorganic components of nanobiofertilizers aid in nitrogen fixation, phosphate solubilization, and soil nutrient replenishment. The nanoparticle coating facilitates a slow and controlled release of nutrients tailored to crop requirements. Nanoencapsulation technology is considered versatile, preserving the components of biofertilizers containing plant growth-promoting rhizobacteria (PGPR), improving their distribution and longevity in fertilizer formulations, and enabling controlled release of PGPR (Vejan et al. 2016). Utilizing nanomaterials in the nano-encapsulation of organic nutrients like zeolite, polymers, and chitosan allows for a gradual and consistent nutrient release to crops. Additionally, the enhanced surface area, nanosize, and increased reactivity of nanoparticle-coated biofertilizers sustainably provide bioavailable nutrients, enabling proper interface and active nutrient uptake for crop fertilization (El-Ghamry et al. 2018). Nanostructured fertilizers comprising neem cake + PGPR demonstrate the potential to expedite crop seedling germination and effectively supply fixed nutrients, thereby enhancing agricultural productivity in certain members of the Fabaceae family (Rahman and Zhang 2018). Application of nanobiofertilizers boosted pest resistance in tomatoes (Gatahi et al. 2015). Nanobiofertilizers based on specific beneficial bacteria have shown the ability to reduce bacterial and fungal infestation in the rhizosphere of Fabaceae family plants (Dikshit et al. 2013; Mukhopadhyay and De 2014; Gouda et al. 2018). Similarly, titanium nanoparticle nanobiofertilizers enhance the adherence of beneficial bacteria to mustard roots (Mishra and Kumar 2009).

## 6.5 Carbon Nanotubes (CNT)

Multi-walled carbon nanotubes (MWCNTs) have shown the ability to penetrate dense seed coverings and modulate gene expression in tomato, tobacco, and mustard plants, thereby promoting their growth (Khodakovskaya et al. 2012). Recent research has indicated that tomato seeds, when treated with carbon nanotubes, exhibit increased moisture absorption, fostering better growth. The utilization of carbon nanofertilizers has also resulted in a notable increase in weight variation. This suggests the potential for designing novel nutrition delivery systems that capitalize on nanoscale porosity domains present on plant surfaces. Another study introduced a polymeric formulation using PVA-starch as a substrate for the controlled release of carbon nanofibers (CNFs) carrying micronutrients such as Cu and Zn. This substrate, composed of PVA and flour, effectively facilitated the gradual release of Cu and Zn NPs from CNFs, enhancing chickpea plant growth. The gradual release mechanism was enabled by the blend of PVA and starch, augmenting the biodegradability of PVA (Kumar et al. 2018). Beyond these findings, the presence of carbon nanotubes (CNTs) in soil has been observed to influence metabolic processes and elevate plant yields and biomass. In various plants such as radish (*Raphanus sativus*), rapeseed (*Brassica napus*), rye, lettuce, maize, and cucumber, CNTs have proven to promote seed germination and subsequent growth. Furthermore, multi-walled carbon nanotubes have shown beneficial effects on the germination of *Brassica juncea* L. and *Phaseolus mungo* L. seeds.

## 6.6 Nanoformulations and Their Smart Delivery Systems

The formulation of any nanofertilizer (NF) should possess desired properties such as high solubility, stability, effectiveness, slow-release pattern, enhanced targeted activity with an effective concentration, and less toxicity for ecology, along with a safe and easy way of delivery and disposal (Tsuji 2001; Boehm et al. 2003; Green and Beestman 2007; Torney et al. 2007). Nanoparticles have significant potential in delivering nutrients to specific target sites within bio-systems. Nutrient loading onto nanoparticles commonly occurs through (a) adsorption onto nanoparticles, (b) attachment on nanoparticles mediated by ligands, (c) encapsulation in nanoparticulate polymeric shells, (d) entrapment of polymeric nanoparticles, and (e) synthesis of nanoparticles composed of the nutrient itself. The mode of fertilizer application significantly impacts its efficiency and its influence on plant systems (Fig. 6.2). Various methods can be employed for delivering nanofertilizers to plants.

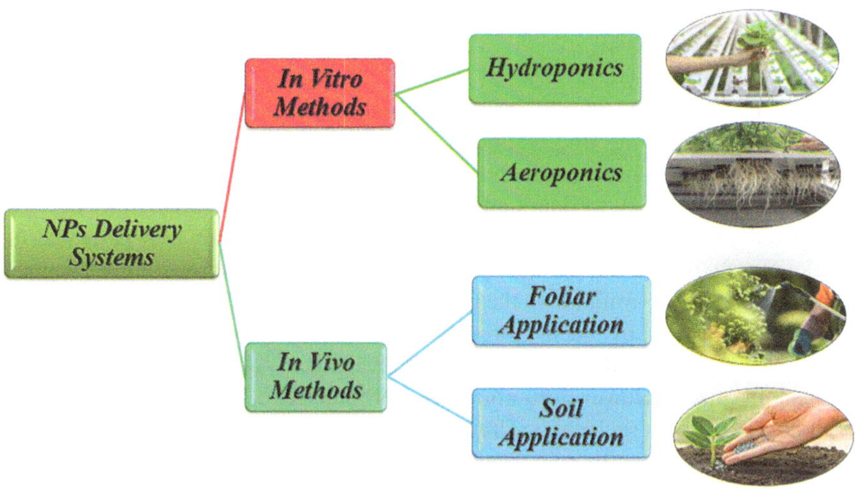


**Fig. 6.2** Different modes of nanoformulation application

## 6.6.1 In Vitro Methods

### 6.6.1.1 Hydroponics

First introduced by Gericke (1937) for dissolved inorganic salts, hydroponics, also referred to as "solution culture," involves growing plants with their roots submerged in a liquid nutrient solution, without soil. Parameters such as nutrient solution volume, maintenance of oxygen levels, and pH require close attention when employing this method for nutrient delivery. In some commercial applications, supporting materials like sand or gravel are utilized. Here, the nutrient solution is flushed from one end while old solution is removed from the other. Drawbacks of hydroponics include frequent susceptibility to pathogens and high moisture levels that might lead to excessive wilting in soil-based plants.

### 6.6.1.2 Aeroponics

Initially reported by Weathers and Zobel (1992), this method involves suspending plant roots in the air while continuously spraying them with a nutrient solution. Aeroponics allows for control over the gaseous environment surrounding the roots. However, sustaining rapid plant growth requires a high level of nutrients, which limits the widespread use of aeroponics.

## 6.6.2 In Vivo Methods

### 6.6.2.1 Foliar Spraying

This method involves directly spraying liquid fertilizers onto leaves, commonly employed for supplying trace elements. Foliar application effectively reduces the time lapse between application and plant uptake, especially during rapid growth phases. It serves as a workaround for restricted nutrient uptake from the soil. Compared to soil application where nutrients are adsorbed on soil particles, the uptake of iron, manganese, and copper can be more efficient through foliar application, as they are less available to the root system (Taiz and Zeiger 2010). Given that nutrient uptake primarily occurs through stomata and leaf epidermal cells, the foliar application method could offer agronomic advantages, particularly when employing nanofertilizers. However, to minimize leaf damage, a standardized application protocol is crucial in such cases. Drawbacks of this method include the necessity of specific spraying times (morning and evening) when stomata are open. Additionally, incorrect concentrations of chemicals (fertilizers) during application could potentially harm the plants.

#### 6.6.2.2 Soil Application

The application of nutrients via soil remains the most prevalent method, utilizing both chemical and organic fertilizers. Several considerations come into play when opting for this method of fertilizer application, including the duration of fertilizer efficacy in the soil, soil texture, salinity, plant sensitivity to salts, salt concentration, and the amendment's impact on soil pH. It is widely recognized that negatively charged soil particles can affect the adsorption of mineral nutrients. Agricultural soils typically exhibit a lower anion exchange capacity compared to cation exchange capacity. Among anions, nitrate retains mobility in the soil solution and is prone to leaching through water movement in the soil. Phosphate ions tend to bind to soil particles containing aluminum or iron due to the exchange of phosphate with the positively charged $Fe^{2+}$, $Fe^{3+}$, and $Al^{3+}$ carrying OH- groups. Consequently, phosphate can become tightly bound, limiting its mobility and availability in the soil, thereby impacting plant growth (Taiz and Zeiger 2010).

## 6.7 Advantages and Opportunities of Nanofertilizers

Nanofertilizers represent a groundbreaking addition to agricultural practices, offering numerous advantages over traditional chemical fertilizers. They hold significant promise for enhancing soil fertility, plant health, and productivity, diverging from conventional methods that often result in nutrient pollution and environmental damage. One of the primary advantages of nanofertilizers lies in their superior ability to deliver nutrients directly to plant cells. This targeted delivery mechanism ensures that plants receive essential nutrients without excess fertilizer runoff, preventing ecological harm to water bodies caused by phosphorus or nitrogen overload. The increased efficacy of nanotechnology-based solutions not only conserves resources but also reduces costs for farmers while minimizing environmental impact on a larger scale. Moreover, nanofertilizer technology effectively addresses the challenge of low micronutrient concentrations, like zinc or iron, crucial for optimal crop yields. These formulations offer slow-release mechanisms that mitigate leaching risks, ensuring sustained plant nutrition during critical growth stages such as flowering and fruiting, maximizing crop production. Additionally, nanobiofertilizers (NBFs) combined with nanofertilizers exhibit the potential to enhance plant stress tolerance. Despite this promise, rigorous assessment and determination of their limitations are imperative before widespread adoption, ensuring judicious applications in crop management (Zulfiqar et al. 2019). The opportunities presented by nanofertilizers in modern agriculture are significant. To harness these opportunities effectively, in-depth research into their properties, modes of action, and effects on plants is essential. Controlling nanofertilizer concentrations is critical, as excessive levels can adversely impact plant growth and yield. Therefore, optimizing dosage quantities specific to different plant types is paramount for practical agricultural use. Further research endeavors should prioritize comprehensive evaluations of nanofertilizers' health implications, effective removal methods, and safe disposal of nanoparticles (NPs). Determining specific dosage recommendations tailored for farmers is crucial for scientific advancements in nanofertilizers, guiding their practical utilization. Additionally, the development and validation of nanofertilizers promise to counterbalance chemical fertilizer production. Policymakers must determine appropriate application schedules and quantities for nanofertilizers, capitalizing on their environmental and economic viability. Overcoming challenges related to raw material availability and scaling up synthesis processes for large-scale distribution is crucial to realizing the potential of nanofertilizers. The success of nanofertilizers hinges on their gradual development into commercially viable and environmentally sustainable products. Research and technological interventions should focus on optimizing manufacturing processes, utilizing cost-effective non-polluting materials, and ensuring biodegradability. Furthermore, accurate methods for identifying and tracking environmental pollutants and comprehensive safety evaluations of nanomaterials are vital, necessitating active funding and collaboration among governmental bodies, corporations, and research groups (Babu et al. 2021; Al-Juthery et al. 2022).

## 6.8 Challenges and Future Outlook

Rising global food demand presents a pressing challenge, amplified by nutrient deficiencies contributing to low crop yields and substantial financial losses in agriculture. While chemical fertilizers have shown short-term benefits, their long-term sustainability is questionable, prompting the pursuit of eco-friendly alternatives in agricultural research. Nanotechnology's emergence in agriculture offers hope through nanotechnology-based smart fertilizers aimed at improving soil health and agricultural output. A comprehensive understanding of their properties is vital for their continuous advancement and commercialization. Streamlining production processes and identifying cost-effective, non-toxic, or biodegradable materials will ensure nanofertilizers' profitability. This discovery bears significant consequences, particularly in terms of environmental sustainability and energy efficiency in global farming practices. For instance, slow-release nanofertilizer solutions could potentially enhance nutrient uptake by up to 40% while reducing water usage by up to 20%. Additionally, decreased reliance on chemical-based fertilizers would curtail carbon emissions linked with agricultural operations upon widespread adoption of improved production practices worldwide. The advent of nanofertilizers is poised to revolutionize sustainable agricultural methods globally, optimizing resource utilization and minimizing reliance on hazardous chemicals detrimental to ecosystems. With steadfast support from various entities, slow-release nanofertilizers have the potential to make a profound impact on the environment, energy consumption, and economic growth worldwide.

## 6.9 Conclusion

The incorporation of nanofertilizers in agriculture presents a vast potential to revolutionize farming, enhancing efficiency and sustainability. Yet, crucial steps must be taken to fully leverage their benefits. Addressing safety concerns, navigating complex regulatory frameworks, and ensuring cost-effectiveness are pivotal in harnessing the complete potential of nanofertilizers for global agricultural advancement. These measures are fundamental in ensuring that this innovative technology can effectively contribute to sustainable and productive farming practices on a broader scale.

## References

Abdelmigid HM, Morsi MM, Hussien NA, Alyamani AA, Alhuthal NA, Albukhaty S (2022) Green synthesis of phosphorous-containing hydroxyapatite nanoparticles (nHAP) as a novel nano-fertilizer: preliminary assessment on pomegranate (Punica granatum L.). Nanomaterials 12:1527

Abobatta WF (2018) Over view of nano-fertilizers. Asian J Ethnopharmacol Med Foods 4:17–20

Adhikary S, Mandal N, Rakshit R, Das A, Kumar V, Kumari N, Choudhary SK, Homa F (2020a) Field evaluation of zincated nanoclay polymer composite (ZNCPC): impact on DTPA-extractable Zn, sequential Zn fractions and apparent Zn recovery under rice rhizosphere. Soil Tillage Res 201:100–110

Adhikary S, Mandal N, Rakshit R, Das A, Kumar V, Kumari N (2020b) Comparative evaluation of nano Zn carriers with conventional Zn sources in terms of release kinetics of Zn and Fe in a laboratory incubation experiment under Inceptisol and Alfisol. J Plant Nutr 43:1968–1979

Afify RR, El-Nwehy SS, Bakry AB, Abd El-Aziz ME (2019) Response of peanut (Arachis hypogaea L.) crop grown on newly reclaimed sandy soil to foliar application of potassium nano-fertilizer. Middle East. J Appl Sci 9(1):78–85 https://doi.org/10.1016/j.theriogenology.2007.04.052

Al-Juthery HW, Lahmoud NR, Alhasan A, Al-Jassani N, Houria A (2022) Nano-fertilizers as a novel technique for maximum yield in wheat biofortification (article review). IOP Conf Ser Earth Environ Sci 1060:012043

Armin M, Akbari S, Mashhadi S (2014) Effect of time and concentration of nano-Fe foliar application on yield and yield components of wheat. Int J Biosci 4:69–75

Askary M, Amirjani MR, Saberi T (2017) Comparison of the effects of nano-iron fertilizer with iron-chelate on growth parameters and some biochemical properties of Catharanthus roseus. J Plant Nutrit 40:974–982

Attia ANE, El-Hendi MH, Hamoda SAF, El-Sayed OS (2016) Effect of nano-fertilizer (Lithovit) and potassium on leaves chemical composition of Egyptian cotton under different planting dates. Int J Plant Product 7:935–942

Avila-Quezada G, Ingle A, Goli'nska, P., & Rai, M. (2022) Strategic applications of nano-fertilizers for sustainable agriculture: benefits and bottlenecks. Nanotechnol Rev 11:2123–2140

Awomi TA, Singh AK, Kumar M, Bordoloi LJ (2012) Effect of phosphorus, molybdenum and cobalt nutrition on yield and quality of mungbean (Vigna radiata L.) in acidic soil of Northeast India. Indian J Hill Farming 25(2):22–26

Babu S, Singh RP, Yadav D, Rathore SS, Raj R, Avasthe R, Yadav SK, Das A, Yadav V, Yadav B (2021) Nanofertilizers for agricultural and environmental sustainability. Chemosphere 292:133451

Boehm AL, Martinon I, Zerrouk R, Rump E, Fessi H (2003) Nanoprecipitation technique for the encapsulation of agrochemical active ingredients. J Microencapsul 20:433–441

Brady NR, Weil RR (1999) In: Brady NR, Weil RR (eds) The nature and properties of soils. Prentice Hall, Upper Saddle River, NJ, pp 415–473
Da-Costa MVJ, Sharma PK (2016) Effect of copper oxide nanoparticles on growth, morphology, photosynthesis, and antioxidant response in Oryza sativa. Photosynthetica 54(1):110–119
Davarpanah S, Tehranifar A, Abadía J, Val J, Davarynejad G, Aran M, Khorassani R (2018) Foliar calcium fertilization reduces fruit cracking in pomegranate (Punica granatum cv. Ardestani). Sci Hortic 230:86–91
De-Bashan LE, Bashan Y (2004) Recent advances in removing phosphorus from wastewater and its future use as fertilizer (1997–2003). Water Res 38:4222–4246
Delfani M, Baradarn Firouzabadi M, Farrokhi N, Makarian H (2014) Some physiological responses of black-eyed pea to iron and magnesium nanofertilizers. Commun Soil Sci Plant Anal 45:530–540
De-Rosa MC, Monreal C, Schnitzer M, Walsh R, Sultan Y (2010) Nanotechnology in fertilizers. Nat Nanotechnol 5:91–91
De-Silva M, Siriwardena DP, Sandaruwan C, Priyadarshana G, Karunaratne V, Kottegoda N (2020) Urea-silica nanohybrids with potential applications for slow and precise release of nitrogen. Mater Lett 272:127839
Dhlamini B, Paumo HK, Katata-Seru L, Kutu FR (2020) Sulphate-supplemented NPK nanofertilizer and its effect on maize growth. Mater Res Express 7:095011
Dikshit A, Shukla SK, Mishra RK (2013) Exploring nanomaterials with PGPR in current agricultural scenario. Lap Lambert Academic Publishing, Saarbrucken
Dimkpa CO, Singh U, Adisa IO, Bindraban PS, Elmer WH, Gardea-Torresdey JL, White JC (2018) Effects of manganese nanoparticle exposure on nutrient acquisition in wheat (Triticum aestivum L.). Agronomy 8:158
Dimkpa CO, Singh U, Adisa IO, Bindraban PS, Elmer WH, Gardea-Torresdey JL, White JC (2018) Effects of Manganese Nanoparticle Exposure on Nutrient Acquisition in Wheat (Triticum aestivum L.). Agronomy 8(9):158. https://doi.org/10.3390/agronomy8090158
Dubey A, Mailapalli DR (2016) Nanofertilisers, nanopesticides, nanosensors of pest and nanotoxicity in agriculture. In: Sustainable agriculture reviews. Springer, pp 307–330
Dubey AN, Chattopadhyaya N, Mandal N (2021) Variation in soil microbial population and soil enzymatic activities under zincated nanoclay polymer composites (ZNCPCs), nano-ZnO and Zn solubilizers in rice rhizosphere. Agric Res 10:21–31
El-Ghamry A, Mosa AA, Alshaal T, El-Ramady H (2018) Nanofertilizers vs. biofertilizers: new insights. Environ Biodivers Soil Secur 2:51–72
Elmer W, White JC (2016) The use of metallic oxide nanoparticles to enhance growth of tomatoes and eggplants in disease infested soil or soilless medium. Environ Sci Nanotechnol 3:1072–1079
Faizan M, Faraz A, Yusuf M, Khan ST, Hayat S (2018) Zinc oxide nanoparticle-mediated changes in photosynthetic efficiency and antioxidant system of tomato plants. Photosynthetica 56:678–686
Farnia A, Ghorbani A (2014) Effect of K nano-fertilizer and N bio-fertilizer on yield and yield components of red bean (Phaseolus vulgaris L.). Biosci Int J 6655:296–303
Gaiotti F, Lucchetta M, Rodegher G, Lorenzoni D, Longo E, Boselli E, Cesco S, Belfiore N, Lovat L, Delgado-L'opez JM, Carmona FJ, Guagliardi A, Masciocchi N, Pii Y (2021) Urea-doped calcium phosphate nanoparticles as sustainable nitrogen nanofertilizers for viticulture: implications on yield and quality of pinot gris grapevines. Agronomy 11:1026
Gatahi DM, Wanyika H, Kihurani AW, Ateka E, Kavoo A (2015) Use of bio-nanocomposites in enhancing bacterial wilt plant resistance, and water conservation in greenhouse farming. In: The JKUAT Scientific Conference on Agricultural Sciences. Technologies and Global Networking, pp 41–52
Gerdini FS (2016) Effect of nano potassium fertilizer on some parchment pumpkin (Cucurbita pepo) morphological and physiological characteristics under drought conditions. Int J Farm Allied Sci 5:367–371
Gericke WF (1937) Hydroponics-crop production in liquid culture media. Science 85:177–178

Ghafariyan MH, Malakouti MJ, Dadpour MR, Stroeve P, Mahmoudi M (2013) Effects of magnetite nanoparticles on soybean chlorophyll. Environ Sci Technol 47:10645–10652
Ghahremani A, Akbari K, Yousefpour M, Ardalani H (2014) Effects of nano-potassium and nano calcium chelated fertilizers on qualitative and quantitative characteristics of Ocimum basilicum. Int J Pharmac Res Schol 3:235–241
Ghormade V, Deshpande MV, Paknikar KM (2011) Perspectives for nano-biotechnology enabled protection and nutrition of plants. Biotechnol Adv 29:792–803
Golbashy M, Sabahi H, Allahdadi I, Nazokdast H, Hossein M (2017) Synthesis of highly intercalated urea-clay nanocomposite via domestic montmorillonite as eco-friendly slow-release fertilizer. Archiv Agron Soil Sci 63:84–95
Gouda S, Kerry RG, Das G, Paramithiotis S, Shin H-S, Patra JK (2018) Revitalization of plant growth promoting rhizobacteria for sustainable development in agriculture. Microbiol Res 206:131–140
Green JM, Beestman GB (2007) Recently patented and commercialized formulation and adjuvant technology. Crop Prot 26:320–327
Gui X, Deng Y, Rui Y, Gao B, Luo W, Chen S, Li X, Liu S, Han Y, Liu L, Xing B (2015) Response difference of transgenic and conventional rice (Oryza sativa) to nanoparticles (γFe2O3). Environ Sci Pollut Res 22:17716–17723
Ha NMC, Nguyen TH, Wang SL, Nguyen AD (2019) Preparation of N:P:K nanofertilizer based on chitosan nanoparticles and its effect on biophysical characteristics and growth of coffee in green house. Res Chem Intermed 45:51–63
Hafeez A, Razzaq A, Mahmood T, Jhanzab HM (2015) Potential of copper nanoparticles to increase growth and yield of wheat. J Nanosci Adv Technol 1:6–11
Hossain KZ, Monreal CM, Sayari A (2008a) Adsorption of urease on PE-MCM-41 and its catalytic effect on hydrolysis of urea. Colloids Surf B: Biointerfaces 62:42–50
Hossain KZ, Sayari A, Monreal CM (2008b) Urease immobilization on pore-expanded mesoporous silica and its catalytic effect on hydrolysis of urea. In: Nonporous materials. World Scientific, pp 697–708
Hu J, Wu C, Ren H, Wang Y, Li J, Huang J (2018) Comparative analysis of physiological impact of γ-Fe2O3 nanoparticles on dicotyledon and monocotyledon. J Nanosci Nanotechnol 18:743–752
Huang J, Xu CC, Ridoutt BG, Wang XC, Ren PA (2017) Nitrogen and phosphorus losses and eutrophication potential associated with fertilizer application to cropland in China. J Clean Product 159:171–179
Hussain HA, Men S, Hussain S, Chen Y, Ali S, Zhang S, Wang L (2019) Interactive effects of drought and heat stresses on morpho-physiological attributes, yield, nutrient uptake and oxidative status in maize hybrids. Sci Rep 9:3890
Hussain N, Bilal M, Iqbal HM (2022) Carbon-based nanomaterials with multipurpose attributes for water treatment: greening the 21st-century nanostructure materials deployment. Biomater Polymer Horiz 1:1–11
Hussein MZ, Zainal Z, Yahaya AH, Foo AFD (2002) Controlled release of a plant growth regulator, α-naphthaleneacetate from the lamella of Zn–Al-layered double hydroxide nanocomposite. J Control Release 82:417–427
Juarez-Maldonado A, Ortega-Ortíz H, Perez-Labrada F, Cadenas-Pliego G, Benavides-Mendoza A (2016) Cu nanoparticles absorbed on chitosan hydrogels positively alter morphological, production, and quality characteristics of tomato. J Appl Bot Food Qual 89:183–189
Kalteh M, Alipour ZT, Ashraf S, Marashi Aliabadi M, Falah Nosratabadi A (2018) Effect of silica nanoparticles on basil (Ocimum basilicum) under salinity stress. J Chem Health Risks 4:49–55
Khalifa NS, Hasaneen MN (2018) The effect of chitosan–PMAA–NPK nanofertilizer on Pisum sativum plants. Biotech 8:193
Khan MR, Rizvi TR (2017) Application of nanofertilizer and nanopesticides for improvements in crop production and protection, in nanoscience and plant-soil systems. Springer, pp 405–427
Khodakovskaya MV, De-Silva K, Biris AS, Dervishi E, Villagarcia H (2012) Carbon nanotubes induce growth enhancement of tobacco cells. ACS Nano 6:2128–2135

Kim JH, Kim D, Seo SM, Kim D (2019) Physiological effects of zero-valent iron nanoparticles in rhizosphere on edible crop, Medicago sativa (alfalfa), grown in soil. Ecotoxicology 28:869–877

Kottegoda N, Siriwardhana D, Priyadarshana W, Sandaruwan C, Madushanka D, Rathnayake U, Gunasekara S, DeAlwis DA, Kumarasinghe A (2014) Compositions and methods for sustained release of agricultural macronutrients. Google Patents

Kottegoda N, Madusanka N, Sandaruwan C (2016) Two new plant nutrient nanocomposites based on urea coated hydroxyapatite: efficacy and plant uptake. Indian J Agric Sci 86:494–499

Kottegoda N, Sandaruwan C, Priyadarshana G, Gunaratne GP, Abeysinghe A, Hettiarachchi S, Karunaratne V, Gehan A (2017) Hydroxyapatite–urea nano-hybrid as efficient plant nutrient systems. ACS Nano 11:1214–1221

Kumar M (2022) Effect of secondary and micronutrients on yield and quality of fox nut (Euryale ferox Salisb.) in subtropical climate. J Plant Nutr 46(7):1–7

Kumar M, Jha AK, Hazarika S, Verma BC, Choudhury BU, Ramesh T, Moirangthem P, Kumar R, Brajendra, Rajkhowa DJ, Kumar A, Devi MH (2016) Micronutrients (B, Zn, Mo) for improving crop production on acidic soils of Northeast India. Natl Acad Sci Lett 39(2):85–89

Kumar R, Ashfaq M, Verma N (2018) Synthesis of novel PVA–starch formulation-supported Cu–Zn nanoparticle carrying carbon nanofibers as a nanofertilizer: controlled release of micronutrients. J Mater Sci 53:7150–7164

Kumar Y, Tiwari KN, Nayak RK, Rai A, Singh SP, Singh AN, Kumar Y, Tomar H, Singh T, Raliya R (2020) Nanofertilizers for increasing nutrient use efficiency, yield and economic returns in important winter season crops of Uttar Pradesh. Indian J Fertilizer 16:772–786

Kumarasinghe AR, Dahanayake D, Karunaratne V, Amaratunga GA (2017) Urea-hydroxyapatite nanohybrids for slow release of nitrogen. ACS Nano 11:1214–1221

León-Silva S, Arrieta-Cortes R, Fernández-Luqueño F, López-Valdez F (2018) Design and production of nanofertilizers. In: Agricultural nanobiotechnology. Springer, Cham, pp 17–31

Li J, Chang PR, Huang J, Wang Y, Yuan H, Ren H (2013) Physiological effects of magnetic iron oxide nanoparticles towards watermelon. J Nanosci Nanotechnol 13:5561–5567

Lin D, Xing B (2007) Phytotoxicity of nanoparticles, inhibition of seed germination and root growth. Environ Pollut 150:243–250

Liu R, Lal R (2014) Synthetic apatite nanoparticles as a phosphorus fertilizer for soybean (Glycine max). Sci Rep 4:5686

Liu R, Lal R (2015) Potentials of engineered nanoparticles as fertilizers for increasing agronomic productions. Sci Total Environ 514:131–139

Mahajan P, Dhoke SK, Khanna AS (2011) Effect of nano-ZnO particle suspension on growth of mung (Vigna radiata) and gram (Cicer arietinum) seedlings using plant agar method. J Nanotechnol 2011:696535

Mandal K, Hati K, Misra A (2009) Biomass yield and energy analysis of soybean production in relation to fertilizer-NPK and organic manure. Biomass Bioenergy 33:1670–1679

Mandal N, Datta SC, Manjaiah KM (2015) Synthesis, characterization and controlled release study of Zn from zincated nanoclay polymer composites (ZNCPCs) in relation to equilibrium water absorbency under zinc deficient Typic Haplustepts. Annals Plant Soil Res 17:187–195

Mandal N, Datta SC, Manjaiah KM, Dwivedi BS, Nain L, Kumar R, Aggarwal P (2018a) Novel chitosan grafted zinc containing nanoclay polymer biocomposite (CZNCPBC): controlled release formulation (CRF) of $Zn^{2+}$. React Funct Polym 127:55–66

Mandal N, Datta SC, Manjaiah KM, Dwivedi BS, Nain L, Kumar R, Aggarwal P (2018b) Zincated nanoclay polymer composites (ZNCPCs): synthesis, characterization, biodegradation and controlled release behaviour in soil. Polym-Plast Technol Eng 57:1760–1770

Mandal N, Datta SC, Manjaiah KM, Dwivedi BS, Kumar R, Aggarwal P (2019) Evaluation of zincated nanoclay polymer composite (ZCNPC) in releasing Zn, P and effect on soil enzyme activities in a wheat rhizosphere. Eur J Soil Sci 70:1164–1182

Mandal N, Datta SC, Manjaiah KM, Dwivedi BS (2021a) Synthesis, characterization, and biodegradation of novel starch grafted zincated nanoclay polymer biocomposites. Polymer Plastics Technol Mater 61:497–515

Mandal N, Datta SC, Dwivedi BS, Manjaiah KM, Meena MC, Bhowmik A (2021b) Zincated nanoclay polymer composite (ZNCPC): effect on DTPA-Zn, olsen-P and soil enzymatic activities in rice rhizosphere. Commun Soil Sci Plant Anal 52:2032–2044

Manimegalai G, Kumar SS, Sharma C (2011) Pesticide mineralization in water using silver nanoparticles. Int J Chem Sci 9:1463–1471

Mastronardi E, Tsae P, Zhang X, Monreal C, Derosa M (2015) Strategic role of nanotechnology in fertilizers: potential and limitations, vol 25–67. Springer, Berlin. https://doi.org/10.1007/978-3-319-14024-7_2

Mendel RR, HaÈnsch R (2002) Molybdoenzymes and molybdenum cofactor in plants. J Exp Bot 53:1689–1698

Mishra VK, Kumar A (2009) Impact of metal nanoparticles on the plant growth promoting rhizobacteria. Digest J Nanomater Biostruct 4:587–592

Morales-Díaz AB, Ortega-Ortíz H, Juárez-Maldonado A, Cadenas-Pliego G, González-Morales S, Benavides-Mendoza A (2017) Application of nanoelements in plant nutrition and its impact in ecosystems. Adv Nat Sci Nanosci Nanotechnol 8:013001. https://doi.org/10.1088/2043-6254/8/1/013001

Mukhopadhyay R, De N (2014) Nano clay polymer composite: synthesis, characterization, properties and application in rainfed agriculture. Global J Bio Biotechnol 3:133–138

Nagaonkar D, Shende S, Rai M (2015) Biosynthesis of copper nanoparticles and its effect on actively dividing cells of mitosis in Allium cepa. Biotechnol Prog 31:557–565

Narendhran S, Rajiv P, Sivaraj R (2016) Influence of zinc oxide nanoparticles on growth of Sesamum indicum L. in zinc deficient soil. Int J Pharm Pharmaceut Sci 8:365–371

Nido PJ, Migo V, Maguyon-Detras MC, Alfafara C (2019) Process optimization potassium nanofertilizer production via ionotropic pre-gelation using alginate-chitosan carrier. In: MATEC Web of Conferences, vol. 268. EDP sciences, 05001. https://doi.org/10.1051/matecconf/201926805001

Pabodha D, Abeywardana L, Sandaruwan C, Herath L, Priyadarshana G (2022) Urea-hydroxyapatite-polymer nanohybrids as seed coatings for enhanced germination. Curr Sci 25(1):50–64

Pan B, Lam SK, Mosier A, Luo Y, Chen D (2016) Ammonia volatilization from synthetic fertilizers and its mitigation strategies: a global synthesis. Agric Ecosyst Environ 232:283–289

Pariona N, Martinez AI, Hdz-García HM, Cruz LA, Hernandez-Valdes A (2017) Effects of hematite and ferrihydrite nanoparticles on germination and growth of maize seedlings. Saudi J Biol Sci 24:1547–1554

Pradhan S, Patra P, Das S, Chandra S, Mitra S, Dey KK, Akbar S, Palit P, Goswami A (2013) Photochemical modulation of biosafe manganese nanoparticles on Vigna radiata: a detailed molecular, biochemical, and biophysical study. Environ Sci Technol 47:13122–13131

Pradhan S, Patra P, Mitra S, Dey KK, Basu S, Chandra S, Palit P, Goswami A (2015) Copper nanoparticle (CuNP) nanochain arrays with a reduced toxicity response: a biophysical and biochemical outlook on Vigna radiate. J Agric Food Chem 63(10):2606–2617

Prasad T, Sudhakar P, Sreenivasulu Y, Latha P, Munaswamy V, Reddy KR, Sreeprasad T, Sajanlal P, Pradeep T (2012) Effect of nanoscale zinc oxide particles on the germination, growth and yield of peanut. J Plant Nutr 35(6):905–927

Preetha PS, Balakrishnan N (2017) A review of nano fertilizers and their use and functions in Soil. Int J Curr Microbiol App Sci 6(12):3117–3133

Raguraj S, Wijayathunga WM, Gunaratne GP, Amali RK, Priyadarshana G, Sandaruwan C, Karunaratne V, Hettiarachchi LS, Kottegoda N (2020) Urea–hydroxyapatite nanohybrid as an efficient nutrient source in Camellia sinensis (L.) Kuntze (tea). J Plant Nutrit 43:2383–2394

Ragab GA, Saad-Allah KM (2020) Seed priming with greenly synthesized sulfur nanoparticles enhances antioxidative defense machinery and restricts oxidative injury under manganese stress in Helianthus annuus (L.) seedlings. J Plant Growth Regul 40(5):1894. https://doi.org/10.1007/s00344-020-10240-y

Rahman KMA, Zhang D (2018) Effects of fertilizer broadcasting on the excessive use of inorganic fertilizers and environmental sustainability. Sustainability 10:759. https://doi.org/10.3390/su10030759

Rajonee AA, Nigar F, Ahmed S, Huq SI (2016) Synthesis of nitrogen nano fertilizer and its efficacy. Can J Pure Appl Sci 10:3913–3919

Rajput VD, Minkina T, Suskova S, Mandzhieva S, Tsitsuashvili V, Chapligin V, Fedorenko A (2018) Effects of copper nanoparticles (CuO NPs) on crop plants: a mini review. Bio Nano Sci 8:36–42

Ramesh M, Palanisamy K, Babu K, Sharma NK (2014) Effects of bulk and nano-titanium dioxide and zinc oxide on physio-morphological changes in Triticum aestivum Linn. J Glob Biosci 3:415–422

Rangaraj S, Gopalu K, Rathinam Y, Periasamy P, Venkatachalam R, Narayanasamy K (2014) Effect of silica nanoparticles on microbial biomass and silica availability in maize rhizosphere. Biotechnol Appl Biochem 61:668–675

Ranjbar S, Rahemi M, Ramezanian A (2018) Comparison of nano-calcium and calcium chloride spray on postharvest quality and cell wall enzymes activity in apple cv. Red Delicious. Sci Hortic 240:57–64

Rathnaweera DN, Pabodha D, Sandaruwan C, Priyadarshana G, Deraniyagala S, Kottegoda N (2019) Urea modified calcium carbonate nanohybrids as a next generation fertilizer. Kotelawala Defence University, Sri Lanka

Rawat S, Pullagurala VL, Hernandez-Molina M, Sun Y, Niu G, Hernandez-Viezcas JA, Peralta-Videa JR, Gardea-Torresdey JL (2018) Impacts of copper oxide nanoparticles on bell pepper (Capsicum annum L.) plants: a full life cycle study. Environ Sci: Nano 5:83–95

Rokana S, Mandal N, Singh M, Pradhan AK, Ghosh M, Sadhu S, Kumari S, Kumar R (2022) Chemical and biological synthesis, characterization and control release pattern of iron nanoparticles (Fe NPs) under incubation experiment in Typic Haplustepts and Typic ustifluvents soil. Clay Res 41:39–59

Rokana S, Mandal N, Singh M, Pradhan AK, Ghosh M, Sadhu S, Kumar V (2023) Synthesis, characterization and controlled release pattern of novel Fe-containing nanoclay polymer composites (Fe-NCPC) in non-calcareous and calcareous soil. J Soil Sci Plant Nutr 23:6139–6154

Rostamizadeh E, Iranbakhsh A, Majd A, Arbabian S, Mehregan I (2020) Green synthesis of Fe2O3 nanoparticles using fruit extract of Cornus mas L. and its growth-promoting roles in barley. J Nanostructure Chem 10:125–130

Rui M, Ma C, Hao Y, Guo J, Rui Y, Tang X, Zhao Q, Fan X, Zhang Z, Hou T, Zhu S (2016) Iron oxide nanoparticles as a potential iron fertilizer for peanut (Arachis hypogaea). Front Plant Sci 7:815

Saharan V, Kumaraswamy RV, Choudhary RC, Kumari S, Pal A, Raliya R, Biswas P (2016) Cu-Chitosan nanoparticle mediated sustainable approach to enhance seedling growth in maize by mobilizing reserved food. J Agric Food Chem 64:6148–6155

Samavini R, Sandaruwan C, De Silva M, Priyadarshana G, Kottegoda N, Karunaratne V (2018) Effect of citric acid surface modification on solubility of hydroxyapatite nanoparticles. J Agric Food Chem 66:3330–3337

Sathiyabama M, Manikandan A (2018) Application of copper-chitosan nanoparticles stimulate growth and induce resistance in finger millet (Eleusine coracana Gaertn.) plants against blast disease. J Agric Food Chem 66:1784–1790

Satdev, Mandal N, Kumar R, Parewa HP, Kumari S, Sadhu S (2022) Different dose of synthesized nanoscale Fe fertilizer application through foliar and soil in rice crop under Typic haplustats. Climate Change Environ Sustain 10:173–191

Savithramma N, Ankanna S, Bhumi G (2012) Effect of nanoparticles on seed germination and seedling growth of Boswellia ovalifoliolata - an endemic and endangered medicinal tree taxon. Nano Vision 2(2):61–68

Sedghi M, Hadi M, Toluie SG (2013) Effect of nano zinc oxide on the germination parameters of soybean seeds under drought stress. Ann West Univ Timisoara Ser Biol 16(2):73

Selva Preetha P, Subramanian KS, Sharmila Rahale C (2014) Sorption characteristics of nanozeolite based slow release sulphur fertilizer. Int J Dev Res 4:225–228
Seydmohammadi Z, Roein Z, Rezvanipour S (2020) Accelerating the growth and flowering of Eustoma grandiflorum by foliar application of nano-ZnO and nano- CaCO3. Plant Physiol Rep 25:140–148
Shah V, Belozerova I (2009) Influence of metal nanoparticles on the soil microbial community and germination of lettuce seeds. Water Air Soil Pollut 197:143–148
Shende S, Rathod D, Gade A, Rai M (2017) Biogenic copper nanoparticles promote the growth of pigeon pea (Cajanus cajan L.). IET Nanobiotechnol 11:773–781
Sheykhbaglou R, Sedghi M, Fathi-Achachlouie B (2018) The effect of ferrous nano-oxide particles on physiological traits and nutritional compounds of soybean (Glycine max L.) Seed. An Acad Bras Ciênc 90(1):485–494
Shinde S, Paralikar P, Ingle AP, Rai M (2018) Promotion of seed germination and seedling growth of Zea mays by magnesium hydroxide nanoparticles synthesized by the filtrate from Aspergillus niger. Arab J Chem 13:3172–3182
Siriwardena T, Priyadarshana G, Sadaruwan C, De-Silva M, Kottegoda N (2017) Urea modified silica nanoparticles: next generation slow release plant nutrients. Third biennial international symposium on polymer science and technology. http://dr.lib.sjp.ac.lk/handle/123456789/7596
Solanki P, Bhargava A, Chhipa H, Jain N, Panwar J (2015) Nano-fertilizers and their smart delivery system. In: Nanotechnologies in food and agriculture. Springer, Cham, pp 81–101
Sood R (2021) Nano-biofertilizers for sustainable agriculture. Just Agric—Multidiscip E-Newsl 1:1–6
Sooyeon L, Sunghyun K, Saeyeon K, Insook L (2013) Assessment of phytotoxicity of ZnO NPs on medicinal plant Fagupyrum esculentum. Environ Sci Pollut Res Int 20:848–854
Suresh S, Karthikeyan S, Jayamoorthy K (2016) Effect of bulk and nano-Fe2O3 particles on peanut plant leaves studied by Fourier transform infrared spectral studies. J Adv Res 7:739–747
Suriyaprabha R, Karunakaran G, Yuvakkumar R, Prabu P, Rajendran V, Kannan N (2012) Growth and physiological responses of maize (Zea mays L.) to porous silica nanoparticles in soil. J Nanopart Res 14:1294
Taheri M, Qarache HA, Qarache AA, Yoosefi M (2016) The effects of zinc-oxide nanoparticles on growth parameters of corn (SC704). STEM Fellowship J 1(2):17–20
Tarafdar JC, Raliya R (2013) Rapid, low-cost, and ecofriendly approach for iron nanoparticle synthesis using aspergillus oryzae TFR9. J Nanoparticles. https://doi.org/10.1155/2013/141274
Tarafdar JC, Raliya R, Rathore I (2012) Microbial synthesis of phosphorous nanoparticle from tri-calcium phosphate using Aspergillus tubingensis TFR-5. J Bionanoscience 6:2. https://doi.org/10.1166/jbns.2012.1077
Tarafdar JC, Raliya R, Mahawar H, Rathore I (2014) Development of zinc nanofertilizer to enhance crop production in pearl millet (Pennisetum americanum). Agric Res 3:257–262
Taran NY, Gonchar OM, Lopatko KG, Batsmanova LM, Patyka MV, Volkogon MV (2014) The effect of colloidal solution of molybdenum nanoparticles on the microbial composition in rhizosphere of Cicer arietinum L. Nanoscale Res Lett 9(1):289
Taran N, Storozhenko V, Svietlova N, Batsmanova L, Shvartau V, Kovalenko M (2017) Effect of zinc and copper nanoparticles on drought resistance of wheat seedlings. Nanoscale Res Lett 12(1):60
Taiz L, Zeiger E (2010) Plant physiology, 5th edn. Sinauer Associates Inc., p 781
Thirunavukkarasu M, Subramanian KS, Kannan P, Balaji T (2018) Response of nano-sulphur to the groundnut. Int J Chem Stud 6:2067–2072
Torney F, Trewyn BG, Lin VSY, Wang K (2007) Mesoporous silica nanoparticles deliver DNA and chemicals into plants. Nat Nanotechnol 2:295–300
Tsuji K (2001) Microencapsulation of pesticides and their improved handling safety. J Microencapsul 18:137–147
Valle SF, Giroto AS, Klaic R, Guimarães GG, Ribeiro C (2019) Sulfur fertilizer based on inverse vulcanization process with soybean oil. Polym Degrad Stabil 162:102–105

Vejan P, Abdullah R, Khadiran T, Ismail S, Boyce AN (2016) Role of plant growth promoting rhizobacteria in agricultural sustainability-A review. Molecules 21:1–17

Venkatachalam P, Priyanka N, Manikandan K, Ganeshbabu I, Indiraarulselvi P, Geetha N, Muralikrishna K, Bhattacharya RC, Tiwari M, Sharma N, Sahi SV (2017) Enhanced plant growth promoting role of phycomolecules coated zinc oxide nanoparticles with P supplementation in cotton (Gossypium hirsutum L.). Plant Physiol Biochem 110:118–127

Wang P, Menzies NW, Lombi E, McKenna BA, Johannessen B, Glover CJ, Kappen P, Kopittke PM (2013) Fate of ZnO nanoparticles in soils and cowpea (Vigna unguiculata). Environ Sci Technol 47:13822–13830

Wang F, Liu X, Shi Z, Tong R, Adams CA, Shi X (2016) Arbuscular mycorrhizae alleviate negative effects of zinc oxide nanoparticle and zinc accumulation in maize plants: A soil microcosm experiment. Chemosphere 147:88–97

Weathers PJ, Zobel RW (1992) Aeroponics for the culture of organisms, tissues and cells. Biotechnol Adv 10:93–115

Yasmeen F, Raja NI, Razzaq A, Komatsu S (2017) Proteomic and physiological analyses of wheat seeds exposed to copper and iron nanoparticles. Biochim Biophys Acta Protein Proteonomics 1865:28–42

Yusuf M, Fariduddin Q, Hayat S, Ahmad A (2011) Nickel: an overview of uptake, essentiality and toxicity in plants. Bull Environ Contam Toxicol 86:1–17

Zhao L, Peralta-Videa JR, Rico CM, Hernandez-Viezcas JA, Sun Y, Niu G, Servin A, Nunez JE, Duarte-Gardea M, Gardea-Torresdey JL (2014) CeO2 and ZnO nanoparticles change the nutritional qualities of cucumber (Cucumis sativus). J Agric Food Chem 62:2752–2759

Zotikova AP, Astafurova TP, Burenina AA, Suchkova SA, Morgalev YN (2018) Morpho-physiological features of wheat (Triticum aestivum L.) seedlings upon exposure to nickel nanoparticles. S-kh Biol 53:578–586. https://doi.org/10.15389/agrobiology.2018.3.578rus

Zulfiqar F, Navarro M, Ashraf M, Munné-Bosch ANA (2019) Nanofertilizer use for sustainable agriculture: advantages and limitations. Plant Sci 289:110270

# Chapter 7
# Regulatory Frameworks, Guidelines, and Policies to Promote Nanofertilizers: Current Status in India and Abroad

**K. S. Subramanian, S. K. Rajkishore, M. Praghadeesh, D. Pradeep, and G. Srinivasan**

**Abstract** Fertilizers are indispensable for sustaining farm productivity. In order to improve fertilizer use efficiencies and minimize the cost of subsidies besides self-reliance in the country, nanofertilizers were introduced in India in 2021 and, thereafter, there is a growing interest in the fertilizer industries to introduce such novel forms of fertilizers with an intent to sustain the productivity of crops with minimal input and without associated environmental hazards. This chapter highlights the status of nanofertilizers, regulatory frameworks available in the country and abroad for notifications, label claims, risk assessment protocols, besides current scenarios of nanofertilizer production and supply and quality control mechanisms. Indeed, India is the first country to have the "*DBT Guidelines for the Evaluation of Nano-based Nano-Agri Inputs and Food Products in India*" since 2020 that facilitated notification of nano urea in 2021 and enabled the commercial scale production and supply of nanofertilizers in the country. The guideline encompasses risk assessment and sequential steps involved in the evaluation of nanoproducts and their safety against various trophic levels such as microbes, predators, parasites, honeybees, earthworms, plants, and humans. After all the tests are completed, nanofertilizers are further evaluated for their compliance with a specific set of stipulated parameters by FCO, namely particle size (hydrodynamic and physical), zeta potential (stability), pH, viscosity, PDI (poly dispersity index) and nutrient concentrations. These parameters are also used for the label claims and the maintenance of the quality

K. S. Subramanian (✉)
Center for Agricultural Nanotechnology, Tamil Nadu Agricultural University, Coimbatore, Tamil Nadu, India

Coromandel Nanotechnology Center, Coimbatore, Tamil Nadu, India
e-mail: kss@tnau.ac.in; kss@coromandel.murugappa.com

S. K. Rajkishore
Department of Environmental Sciences, Coimbatore, Tamil Nadu, India

M. Praghadeesh · D. Pradeep · G. Srinivasan
Center for Agricultural Nanotechnology, Tamil Nadu Agricultural University, Coimbatore, Tamil Nadu, India

 
A. K. Patra (ed.), *Nanofertilizers in Agriculture*, Sustainability Sciences in Asia and Africa, https://doi.org/10.1007/978-981-97-8112-6_7

control by the industries. Further, there is a DST set of guidelines to meet the safety requirements of industries undertaking the production and supply of nanoproducts, including nanofertilizers. Overall, this book chapter concludes that there is a set of stipulated stringent guidelines for the evaluation, notification, production, and supply of nanofertilizers and the same is being used for the quality control of nanofertilizers.

**Keywords** Nanomaterials · Agriculture · Biosafety tests · Regulatory frameworks · Nano-fertilizers · Nanotech policy

## 7.1 Status of Nanofertilizers

Indian agriculture has undergone metamorphosis that resulted in paradigm shift from conventional to precision agriculture in order to address a wide array of challenges encompassing burgeoning population, shrinking farm land, restricted water availability, imbalanced crop nutrition, multi-nutrient deficiencies, crop yield barriers and decline in soil organic matter. Nanotechnology deals with atomic manipulations that enable to evolve processes and products which are precise to deliver inputs in agriculture without associated environmental hazard. Nanotechnology applications in agriculture include diagnostic devices for early detection of plant diseases, nanoagricultural inputs such as nanofertilizer (Subramanian et al. 2015; Subramanian and Rajkishore 2018; Nongbet et al. 2022; Verma et al. 2022), nanoherbicides (Abigail and Chidambaram 2017; Forini et al. 2022) and nanoinsecticides (Wang et al. 2022), nanoseed science, plant health management (Subramanian et al. 2016), nanofood systems (Anusuya et al. 2016; Jincy et al. 2017; Jan et al. 2023), besides environmental remediation (Rajkishore et al. 2023).

Nanotechnology is being exploited to develop nano-enabled processes and agri-inputs intended to enhance use efficiencies, reduce the cost, use of minimal inputs besides preserving natural environment. Among the various nano-agri inputs, nanofertilizers were given a national priority as fertilizers are an indispensable input in agriculture. Over the past seven decades, the food grain production of India has reached the all-time high of **325 million tonnes** in 2022–2023 from just **50.8 million tonnes** in 1950s (Fig. 7.1). The 6.5-fold increase in food grain production is closely commensurate with the increase in fertilizer nutrient consumption from **1.11 million tonnes** to **43.5 million tonnes** during the same reported period. In order to sustain the agricultural productivity over the years, the country subsidized the fertilizers to the tune of **Rs. 2,25,000 Crores** (2.25 trillion INR) during 2022—2023, which is a huge exchequer for the country. The Government of India

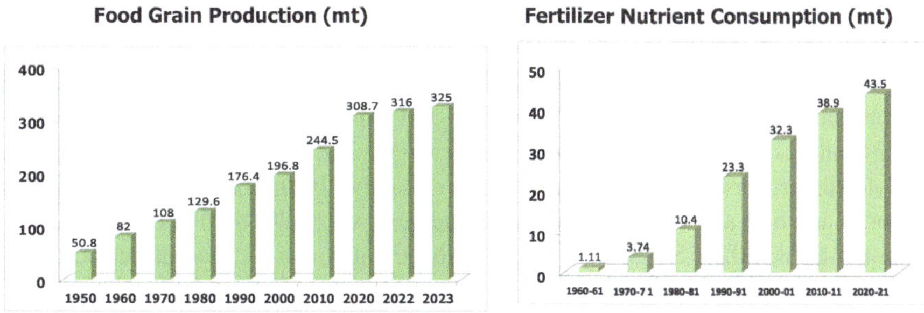


**Fig. 7.1** Food grain production and fertilizer nutrient consumption over the years

widely believes that the introduction of nanofertilizers can substantially reduce the burden of fertilizer subsidy.

Nanofertilizers are known to improve the nutrient use efficiencies due to their unique property of high surface mass ratio. Nanofertilizers are synthesized by **top down** (size reduction by physical method) or **bottom up** (building of atoms by chemical or biological methods). In physical methods, the carrier particles are ground using high energy ball milling, and nutrients are fortified singly or in combinations with carriers in nano-dimensions. Clays are used as carriers for developing nanofertilizers by loading cationic nutrients ($NH_4^+$, $K^+$, $Ca^{++}$, $Mg^{++}$) as it is and anionic nutrients ($NO_3^-$, $PO_4^-$, $SO_4^-$) after suitable surface modifications.

The nanofertilizers developed using a top-down approach are reported to improve the nutrient use efficiencies, reduce the quantity and fertilizer use, besides enhancing productivity of crops with reference to nitrogen (Subramanian and Rahale 2012; Manikandan and Subramanian 2014; Subramanian et al. 2015, Subramanian and Rajkishore 2018; Bernela et al. 2021; Latha et al. 2023), phosphorous (Bansiwal et al. 2006; Bernela et al. 2021), potassium (Subramanian and Rahale 2012), NPK (Kumar et al. 2022), sulfur (Thirunavukkarasu and Subramanian (2014a, b), Ca and Mg (Reddy and Subramanian (2015), zinc (Yuvaraj and Subramanian 2018), and boron (Selva Preetha 2011). Although these methods of developing nanofertilizers helped improve nutrient use efficiencies as a result of sustained release of nutrients, the technology is too expensive due to the intensive energy requirement and mining for raw materials from natural sources. Further, nanofertilizers developed by physical methods are highly vulnerable for agglomeration into microscale particles. Consequently, chemical and biological methods become popular and practically feasible for a scale-up of technology.

The chemical and biological methods are used to enable atomic manipulations that lead to reduction, nucleation and crystal growth, and the formation of stable nanoparticles. Encapsulation of nutrients singly or in combination was done to achieve improved nutrient efficiencies. Natural or synthetic polymers were used for the encapsulation of nutrients. These methods were used to develop slow-release fertilizers carrying nitrogen (Ghafoor et al. 2021; Beig et al. 2020), phosphorous (Bansiwal et al. 2006; Chen et al. 2021), combined N and P as diammonium phosphates (Liu et al. 2021), multi-nutrients (Mohanraj et al. 2022a, b), sulfur (Subramanian et al. 2022), zinc (Yuvaraj and Subramanian 2020; Yuvaraj et al. 2023), and composite of nutrients (Khan et al. 2021; Lateef et al. 2016). The nanofertilizer research outcomes have already been reviewed extensively (Subramanian et al. 2015; Subramanian and Rajkishore 2018; Yuvaraj et al. 2024; Shoukat et al. 2024; Singh et al. 2024; Abd-Elsalam 2024; Madlala et al. 2024; Huang et al. 2024; El-Mahrouk et al. 2024)

The reported literatures on nanofertilizers clearly show that nano-fertilizers have a potential role to play in sustaining farm productivity. Among the approaches employed in the production of nanofertilizers, chemical and biological methods are appropriate as these methods yield a stable nanoformulation and scale-up is possible. All the five notified nanofertilizers in India such as nano urea (IFFCO—2021), nano DAP (Coromandel, IFFCO, Zuari—2023) and Nano-P (TERI—2023) by FCO are manufactured based on the chemical and biological processes. The fertilizer industries are in the process of large-scale production and supply of nanofertilizers. The lead nanofertilizers that are available in India and abroad are listed in Table 7.1.

The literature review on evaluation of commercially viable nanofertilizers suggests that there is a wide array of nanofertilizer products being developed, evaluated, and notified in different countries for large scale production and supply. The countries undertaking nanofertilizer research include India, USA, UK, Spain, Malaysia, Taiwan, Germany, Thailand, China, Japan, and Iran. But India is one of the early birds in developing regulatory protocols for the use of nano-agri inputs and nanofood products (Adholeya and Singh 2020), which led to the notification of five nanofertilizers for commercial production. Indeed, fertilizer industries have started building nanofertilizer manufacturing units in order to meet the targeted fertilizer production.

**Table 7.1** Nanofertilizers commercially available across the world

| No | Country | Company | Commercial products | Contents | Reference |
|---|---|---|---|---|---|
| 1 | India | IFFCO, New Delhi | Nano urea (liquid) | N supplement (4% N) | FCO (2021) |
| 2 | | IFFCO, New Delhi | Nano DAP (liquid) | N and P supplement (8% N; 16% $P_2O_5$) | FCO (2023)[a] |
| 3 | | Coromandel International Ltd., Hyderabad | Gromor nano DAP (liquid) | N and P supplement (2% N; 5% $P_2O_5$) | FCO (2023)[b] |
| 4 | | Zuari Farm Hub Ltd. | Nano DAP (liquid) | N and P supplement (6% N; 16% $P_2O_5$) | FCO (2023)[c] |
| 5 | | | Nano urea (liquid) | N supplement (8% N) | FCO (2023)[d] |
| 6 | | TERI, New Delhi | Nano P (liquid) | P supplement (8% N) | FCO (2023) |
| 7 | | Tropical Agrosystem India, Karnal | NPK, P, Zn, Ca | Proteino-lacto-gluconate chelated with micronutrient, vitamins, probiotics, seaweed extracts, and humic acid | Kah et al. (2018) |
| 8 | | Aditya Birla Nuvo Ltd., Veraval | Neem coated urea | N fertilizer for plants | Patent US9884792B2 |
| 9 | | JU Agri Sciences Pvt. Ltd., Janakpuri, New Delhi | Nano max NPK fertilizer | Multiple organic acids chelated with macro nutrients, amino acids, organic carbon, micronutrients, vitamins, and probiotics | Patel et al. (2020) |
| 10 | | Nano Green Sciences, Inc., Delhi | Nano green | Extract of corn, grain, soybeans, potatoes, coconut, and palm | Rajput et al. (2021) |
| 11 | | Alert Biotech, Nashik | Nano boron | 20% Nano B used as micronutrient | Rajput et al. (2021) |
| 12 | | Shan Maw Myae Trading Co., Ltd., Yangon | Nano micronutrient (EcoStar) | Zn, 6%; B, 2%; Cu, 1%; Fe, 6%+; EDTA Mo, 0.05%; Mn, 5%+; and AMINOS, 5% | Dimkpa and Bindraban (2017) |
| 13 | USA | Agro Nanotechnology Corp., Miami, Florida | Nano-GroTM | Plant growth regulator and immunity enhancer | Jędrszczyk and Ambroszczyk (2016) |
| 14 | | Urth Agriculture, Monterey, CA | Nano-ag answer | Microorganisms, sea kelp and mineral electrolyte | Avila-Quezada et al. (2022) |
| 15 | | Nualgi America, Inc., San Marcos, CA | Nualgi foliar spray | Combining 12 essentials nutrients loaded in nano-silica (henceforth NF-A) | Gil-Díaz et al. (2022) |

(continued)

**Table 7.1** (continued)

| No | Country | Company | Commercial products | Contents | Reference |
|---|---|---|---|---|---|
| 16 | Taiwan | Land Green and Technology Co., Ltd. | NovaLand-Nano | Nano macro- and micro-elements for plant growth (henceforth NF-B) | Gil-Díaz et al. (2022) |
| 17 | | Land Green and Technology Co., Ltd. | Titanium dioxide ($TiO_2$)—Universal pigment (20 nm) | Titanium dioxide 99% | Avila-Quezada et al. (2022) |
| 18 | | Land Green and Technology Co., Ltd. | Silicon dioxide ($SiO_2$)—Universal stabilizer agent (20–60 nm) | Silicon dioxide 99% | Avila-Quezada et al. (2022) |
| 19 | | Land Green and Technology Co., Ltd. | Manganese dioxide ($MnO_2$)—Universal purifier (1-50 nm) | Manganese dioxide 99.9% | Avila-Quezada et al. (2022) |
| 20 | | Land Green and Technology Co., Ltd. | Selenium colloid (Se)—Universal antioxidant (1–20 nm) | Selenium colloid 99.9% | Avila-Quezada et al. (2022) |
| 21 | | Lazuriton Nano Biotechnology Co., Ltd. | Nano—Organic compound fertilizer | Plant growth promoters in vegetables, crops, and flowers | Commission No. 333-210091, USA |
| 22 | | SMTET Eco-technologies Co., Ltd. | Nano ultra—fertilizer | Organic matter, 5.5%; nitrogen, 10%; $P_2O_5$, 9%; $K_2O$, 14%; and MgO, 3% | Dimkpa and Bindraban (2017) |
| 23 | Thailand | Pannaraj Intertrade, Bangkok | Master Nano chitosan organic fertilizer | Water soluble liquid chitosan, organic acid, salicylic acids, and phenolic compounds | Dimkpa and Bindraban (2017) |
| 24 | | The Best International Network Co., Ltd., Bankok | Supplementary powder (nano capsule) | N, 0.5%; $K_2O$, 3.9%; Ca, 2%; Mg, 0.2%; S, 0.75%; $P_2O_5$, 0.7%; Fe, 0.03%; Cu, 0.007%; Zn, 0.004%; Mn, 0.004% | Dimkpa and Bindraban (2017) |
| 25 | | Green Organic World Co., Ltd., Surin | Plant nutrition powder (green nano) | N, 0.5%; $P_2O_5$, 0.7%; $K_2O$, 3.9%; Ca, 2%; Mg, 0.2%; S, 0.8%; Fe, 1.0%; Mn, 49 ppm; Cu, 17 ppm and Zn, 12 ppm | Dimkpa and Bindraban (2017) |
| 26 | Iran | Fanavar Nano-Pazhoohesh Markazi Company, Tehran | Biozar nanofertilizer | Combination of organic materials, micronutrients, and macromolecules | Avila-Quezada et al. (2022) |
| 27 | Japan | Japan | Poly olefin resin-coated urea | N supplements for plants | Tian et al. (2019) |

| | | | | | |
|---|---|---|---|---|---|
| 28 | China | Qingdao Hibong Fertilizer Co., Ltd., Qingdao | Hibong biological fulvic acid | Nanofertilizer, humic acid, chitosan | Avila-Quezada et al. (2022) |
| 29 | | Qingdao Hibong Fertilizer Co., Ltd., Qingdao | Seaweed nano organic carbon fertilizer | NPK: 2-3-3, seaweed extract ≥5%, organic matter: 35%, humic acid ≥5%, amino acid ≥5% | Avila-Quezada et al. (2022) |
| 30 | Spain | Fosvit K30; Kimitec Group | P, K | Fertilizer with high content of P (30%) and K (20%) | Rajput et al. (2021) |
| 31 | United Kingdom | AFME Trading Group | Nano zinc chelate fertilizer | Growth enhancer | Rajput et al. (2021) |
| 32 | | AFME Trading Group | Nano iron and calcium, potassium chelate fertilizer | Plant growth regulator and accelerator | Rajput et al. (2021) |
| 33 | Germany | AC International Network Co., Ltd. | Nano calcium (magic green) | $CaCO_3$, 77.9%; $MgCO_3$, 7.4%; $SiO_2$, 7.47%; K, 0.2%; Na, 0.03%; P, 0.02%; Fe, 7.4 ppm; $Al_2O_3$, 6.3 ppm; Sr, 804 ppm; sulfate, 278 ppm; Ba, 174 ppm; Mn, 172 ppm; and Zn, 10 ppm | Dimkpa and Bindraban (2017) |
| 34 | Malaysia | WAI International Development Co., Ltd., Sungai Buloh | PPC nano | M protein, 19.6%; $Na_2O$, 0.3%; $K_2O$, 2.1%; $(NH_4)\ 2SO_4$, 1.7% and diluent, 76% | Dimkpa and Bindraban (2017) |

## 7.2 Biosafety of Nanofertilizers

Fertilizer production includes synthesis, characterization, and product development. The synthesis of nanofertilizers is achieved by top down or bottom-up approaches. In many cases, nanofertilizers in India are produced by converting the existing bulk fertilizers into nanoforms by dissolving them and the nutrients are then encapsulated into nanosized carriers like chitosan or any other biopolymers. In some cases, microbes are being used to develop nanofertilizers for size reduction and encapsulation. Thus, biosafety of nanofertilizers is to be assessed holistically from the point of process, product, and performance of the crops without associated environmental ill-effects in soil-water-plant-atmospheric continuum. The biosafety of nanofertilizers can be ensured by employing a standard set of protocols to assess the nanotoxicity (Rajkishore et al. 2013). Nanotoxicity studies are undertaken in vitro, in vivo tests, and in silico analysis (Dusinska et al. 2009). In invitro assays, cytotoxicity, and genotoxicity are widely used to evaluate the toxicity of nanofertilizers. Cytotoxicity indicates the loss of some cellular or intercellular structures or functions including lethal toxicity. It measures the potential to cause cell and tissue injury in target cells like skin, lungs, kidneys and liver. Some of these SOPs are built in OECD guidelines that served as the regulatory framework (Dusinska et al. 2012). All the relevant OCED protocols are to be adopted to get the required data set for the registration of nanofertilizers in India or abroad (Table 7.2).

**Table 7.2** OECD protocols for assessing the biosafety of nanofertilizers

| S. no | OECD test no | Name of the test | Data of issue | DBT guideline[a] |
|---|---|---|---|---|
| **Microbes** | | | | |
| 1 | Test no. 217 | Soil microorganisms: Carbon transformation test | 21 Jan 2000 | 9.2 |
| 2 | Test no. 216 | Soil microorganisms: Nitrogen transformation test | 21 Jan 2000 | 9.2 |
| 3 | Test no. 471 | Bacterial reverse mutation test | 21 Jul 1997 | 9.2/11.2.10 |
| **Insects** | | | | |
| 1 | Test no. 214 | Acute contact toxicity test for honeybees | 21 Sep 1998 | 9.3 |
| 2 | Test no. 213 | Acute oral toxicity test for honeybees | 21 Sep 1998 | 9.3 |
| 3 | | Parasitoids (beneficial insects) *Trichogramma pretiosum* and *T. chilonis* | Mohan et al. (2017) | |
| 4 | | Predators (beneficial insects) *Chrysoperla zasrowi sillemi* | Mohan et al. (2020) | |
| **Earthworm** | | | | |
| 1 | Test no. 222 | Earthworm reproduction test (*Eisenia fetida* or *Eisenia andrei*) | 29 Jul 2016 | 9.2/10.3 |
| 2 | Test no. 207 | Acute toxicity tests for earthworms | 04 Apr 1984 | 9.3 |
| 3 | Test no. 317 | Bioaccumulation in terrestrial oligochaetes | 23 Jul 2010 | 10.3 |

(continued)

**Table 7.2** (continued)

| S. no | OECD test no | Name of the test | Data of issue | DBT guideline[a] |
|---|---|---|---|---|
| **Plants** | | | | |
| 1 | Test no. 208 | Terrestrial plant test: Seedling emergence and seedling growth test | 17 Aug 2006 | 9.2 |
| **Aquatic system** | | | | |
| 1 | Test no. 203 | Acute toxicity test in fish | 18 Jun 2019 | 9.2 |
| 2 | Test no. 229 | Short-term reproduction assay for fish | 02 Oct 2012 | 9.2 |
| 3 | Test no. 211 | Daphnia magna reproduction test | 02 Oct 2012 | 9.2/10.3 |
| 4 | Test no. 201 | Freshwater alga and cyanobacteria, growth inhibition test | 28 Jul 2011 | 10.3 |
| 5 | Test no. 202 | Acute immobilisation test in Daphnia sp. | 23 Nov 2004 | 9.2/10.3 |
| 6 | Test no. 305 | Bioaccumulation in fish: Aqueous and dietary exposure | 02 Oct 2012 | 10.3 |
| **Avian/birds** | | | | |
| 1 | Test no. 223 | Oral toxicity test in birds | 29 Jul 2016 | 9.3 |
| 2 | Test no. 206 | Reproduction test in birds | 04 Apr 1984 | 9.3 |
| 3 | Test no. 205 | Dietary toxicity test in birds | 04 Apr 1984 | 9.3 |
| **Rodents** | | | | |
| 1 | Test no. 408 | Repeated dose 90-day oral toxicity study in rodents | 27 Jun 2018 | 12.2.6.2 |
| 2 | Test no. 424 | Neurotoxicity study in rodents | 21 Jul 1997 | 9.3 |
| **Livestock** | | | | |
| 1 | Test no. 505 | Residues in livestock | 25 Jan 2007 | 9.3 |
| 2 | Test no. 503 | Metabolism in livestock | 25 Jan 2007 | 9.3 |
| **Animal system** | | | | |
| 1 | Test no. 431 | In vitro skin corrosion: Reconstructed human epidermis (RHE) test method | 18 Jun 2019 | 9.1 |
| 2 | Test no. 453 | Combined chronic toxicity/carcinogenicity | 27 Jun 2018 | 12.2.6.5 |
| 3 | Test no. 451 | Carcinogenicity | 27 Jun 2018 | 9.3 |
| 4 | Test no. 414 | Prenatal developmental toxicity test | 27 Jun 2018 | 9.3 |
| 5 | Test no. 433 | Acute inhalation toxicity: Fixed concentration procedure | 27 Jun 2018 | 9.1 |
| 6 | Test no. 443 | Extended one-generation reproductive toxicity study | 27 Jun 2018 | 9.3/12.2.6.6 |
| 7 | Test no. 402 | Acute dermal toxicity | 09 Oct 2017 | 12.2.6.2 |
| 8 | Test no. 437 | Bovine corneal opacity and permeability test method for identifying (i) Chemicals inducing serious eye damage and (ii) Chemicals not requiring classification for eye irritation or serious eye damage | 09 Oct 2017 | 9.1 |
| 9 | Test no. 405 | Acute eye irritation/corrosion | 09 Oct 2017 | 9.3 |
| 10 | Test no. 489 | In vivo mammalian alkaline comet assay | 29 Jul 2016 | 9.1 |
| 11 | Test no. 487 | In vitro mammalian cell micronucleus test | 29 Jul 2016 | 9.1/11.2.10 |

(continued)

**Table 7.2** (continued)

| S. no | OECD test no | Name of the test | Data of issue | DBT guideline[a] |
|---|---|---|---|---|
| 12 | Test no. 476 | In vitro mammalian cell gene mutation tests using the *Hprt* and *xprt* genes | 29 Jul 2016 | 9.1/11.2.10 |
| 13 | Test no. 474 | Mammalian erythrocyte micronucleus test | 29 Jul 2016 | 12.2.6.4 |
| 14 | Test no. 473 | In vitro mammalian chromosomal aberration test | 29 Jul 2016 | 9.1/11.2.10 |
| 15 | Test no. 404 | Acute dermal irritation/corrosion | 28 Jul 2015 | 9.3 |
| 16 | Test no. 429 | Skin sensitisation | 23 Jul 2010 | 9.3 |
| 17 | Test no. 417 | Toxicokinetics | 23 Jul 2010 | 9.3 |
| 18 | Test no. 436 | Acute inhalation toxicity—Acute toxic class method | | 9.1 |
| 19 | Test no. 403 | Acute inhalation toxicity | 08 Sep 2009 | 9.3 |
| 20 | Test no. 407 | Repeated dose 28-day oral toxicity study in rodents | 16 Oct 2008 | 9.3 |
| 21 | Test no. 426 | Developmental neurotoxicity study | 15 Oct 2007 | 9.3 |
| 22 | Test no. 428 | Skin absorption: In vitro method | 23 Nov 2004 | 9.1/12.2.5.3 |
| 23 | Test no. 416 | Two-generation reproduction toxicity | 22 Jan 2001 | 9.3 |
| 24 | Test no. 419 | Delayed neurotoxicity of organophosphorus substances: 28-day repeated dose study | 27 Jul 1995 | 9.3 |
| 25 | Test no. 406 | Skin sensitisation | 17 Jul 1992 | 9.3 |
| 26 | Test no. 482 | Genetic toxicology: DNA damage and repair, unscheduled DNA synthesis in mammalian cells in vitro | 23 Oct 1986 | 11.2.10 |
| **General** | | | | |
| 1 | Test no. 318 | Dispersion stability of nanomaterials in simulated environmental media | 09 Oct 2017 | 9.2/11.2.2 |
| 2 | Test no. 307 | Aerobic and anaerobic transformation in soil | 24 Apr 2002 | 9.2 |
| **Human system** | | | | |
| **Cytotoxicity** ATP cell titer-Glo, neutral red uptake, LDH release, MTT, XTT, cell impedance, trypan blue, BrdU, Alamar blue, WST-1, live/dead cell counting, colony-forming efficiency | | | | 9.1.5 |

[a] Guidelines for evaluation of nanobased agri-inputs and food products in India (Adholeya and Singh 2020)

Biosafety of nanofertilizers that are already notified has been evaluated extensively and reported to the Central Fertilizer Committee prior to the notification in FCO. Some of the potential nanoparticles that are being employed in nanofertilizer production have been tested for their biosafety and are reported in Table 7.3.

**Table 7.3** Nanotoxicity of various nanoparticles/products at various trophic levels

| Model system | Nanoproducts | Effects | References |
|---|---|---|---|
| Algae | $TiO_2$ | Affected photosynthetic activity | Navarro et al. (2008), Aruoja et al. (2009) |
| | $CeO_2$ | Inhibited growth | Hoecke et al. (2009) |
| Microbes | ZVI | Cell disruption | Lee et al. (2008) |
| | CuO | Antibacterial activity | Mahapatra et al. (2008); Dimkpa et al. (2011) |
| | Ag | Increased cell permeability and ultimately cell death | Sondi and Salopek-Sondi (2004) |
| | ZnO | 100% mortality | Jiang et al. (2009) |
| Plants | CNT | Programmed cell death | Shen et al. (2010) |
| | ZnO | Inhibited seed germination and root growth | Lin and Xing (2007) |
| Fish | Se | Hyper-accumulation | Li et al. (2008) |
| | ZVI | Morphologic changes | Li et al. (2009a) |
| Rodents | $Al_2O_3$ | Inflammation, crossed blood–brain barrier | Li et al. (2009b) |
| | $TiO_2$ | Inflammation | Oberdörster et al. (1992), Baggs et al. (1997) |
| | ZVI | Toxic to nerve cells | Phenrat et al. (2009) |
| | CNT | Death, necrosis, inflammation, and cell injury | Lam et al. (2004), Shvedova et al. (2010) |
| | Au | Transferred from mother's placenta to fetus | Warheit et al. (2004) |
| Humans | Magnetite | Neurodegenerative diseases | Kirschvink et al. (1992), Dunn et al. (1995) |
| | CNT | Mesothelioma | Donaldson et al. (2006), Fisher et al. (2012) |
| | Au | Penetrated the sperm head and tails causing fragmentation | Wiwanitkit et al. (2009) |

## 7.3 Assays Used for Assessing Safety of Nanofertilizers for Humans

The biosafety of nano-agri inputs or nano-agri products can be evaluated using a set of appropriate assays. In general, nanotoxicity studies include in vitro, in vivo tests, and in silico analysis. Cytotoxic, genotoxic, and alterations in gene expression assays are commonly suggested by regulatory bodies and acceptable across the globe (Rajkishore et al. 2013). The OECD guidelines are commonly adopted for assessing the toxicity of chemicals and the same being employed to determine the biosafety of nanomaterials (OECD 2009) depending on the specificities of nano-products. It is suggested to evaluate nano-agri inputs and nano-agri products as proposed in the DBT draft guidelines (Subramanian and Rajkishore 2018). The cytotoxicity and genotoxicity assays are widely used for assessing the toxicity of nanofertilizers or nanoproducts on humans.

## 7.3.1 *Cytotoxicity*

The cytotoxicity assay is to measure the intensity of toxicity on the normal cell line in order to determine the impact of nanotoxicity. MTT assay is a widely used cytotoxicity test to determine the live cells in relation to dead cells. In this assay, the MTT (3-[4,5-dimethylthiazol-2-yl]-2,5 diphenyl tetrazolium bromide) yellow color is changed to violet color and insoluble formazan crystal with NADPH dependent mitochondrial oxidoreductase enzymes in live cell that reduces NADPH into $NAD^+$, and MTT into formazan crystal, respectively. The insoluble formazan is dissolved in dimethyl sulfoxide or sodium dodecyl sulfate detergent and the chromophore is read between 540 and 620 nm (Riss et al. 2019). The change in color indicates the living status and functionality.

There are 8 assays that are being adopted to measure the cytotoxicity:

(a) Cell titer-Glo luminescent cell viability assay.
(b) Neutral red uptake.
(c) LDH assay.
(d) MTT assay.
(e) XTT assay.
(f) Live and dead cell counting.
(g) WTS-8 assay.
(h) Trypan blue assay.

### 7.3.1.1 Cell Titer-Glo Luminescent Cell Viability Assay

This test enumerates the number of viable cells present in the culture by measuring the ATP, the major source of energy of living cells.

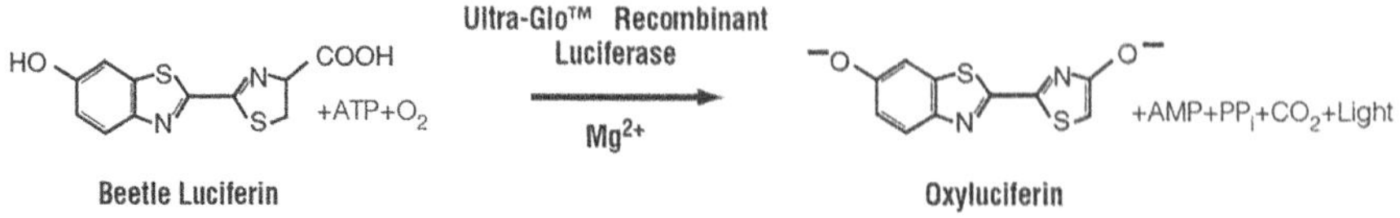


**Cell titer-Glo luminescent cell viability assay. (Courtesy: Hannah et al. 2001)**

In this test, the luciferin substrate is converted by luciferase enzyme in the presence of ATP and $Mg^{2+}$ into a luminescent oxyluciferin molecule. Thus, metabolically active cells produce luminescence, whereas dead cells do not produce it (Hannah et al. 2001).

### 7.3.1.2 Neutral Red Uptake

This assay is employed to measure the viable cells as the ability of neutral red dye to move into lysosome at the optimum physiological pH of the cell. Neutral red is a weak cationic dye, which moves to the lysozome of cell through nonionic passive diffusion mode and binds to the phosphate group of anionic charge present in the lysozomal matrix through electrostatic hydrophobic bonds. After the entry of the dye, a proton gradient occurs in the lysozome to maintain the pH gradient with cytoplasm, during which the dye completely becomes cationic and is retained within the lysosome. If the cell is not alive, there won't be a balance in the pH gradient, such that the dye retention in lysozome will be reduced (Repetto et al. 2008). The amount of dye retained in the cell is measured at 540 nm. The intensity of absorption increases with an increase in the number of viable cells.

#### 7.3.1.3 Lactate Dehydrogenase (LDH) Assay

The LDH test provides the number of dead or unviable cells by measuring the lactate dehydrogenase enzyme produced during the unfavorable condition of cells. LDH enzyme converts the stored form of energy (lactate) into an active energy (pyruvate), and during this conversion the cofactor NAD+ is reduced to NADH.

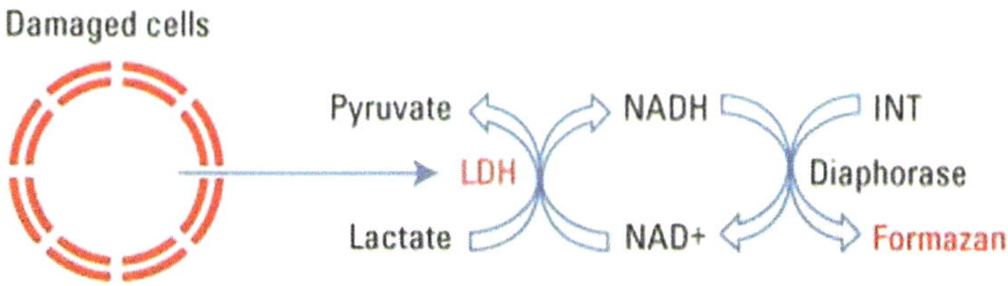


**Lactate dehydrogenase assay. (Courtesy: Parhamifar et al. 2013)**

The diopharase enzyme (added externally) is capable of converting iodonitro tetrazolium salt INT (added externally) into a red colored formazon product by utilizing the proton released during the oxidation of NADH to NAD+ (Parhamifar et al. 2013). The red colored formazon product was measured at 490 nm. The absorbance increases with an increase in the number of nonviable cells and vice versa.

#### 7.3.1.4 MTT Assay

The MTT refers to the chemical name 3-(4,5-dimethylthiazol-2-yl)-2,5-diphenyltetrazolium bromide of the substrate molecule. It is used to determine the number of viable cells by measuring the mitochondrial succinate dehydrogense enzyme which converts succinate to fumarate in Kreb's cycle of respiration (Stockert et al. 2018)

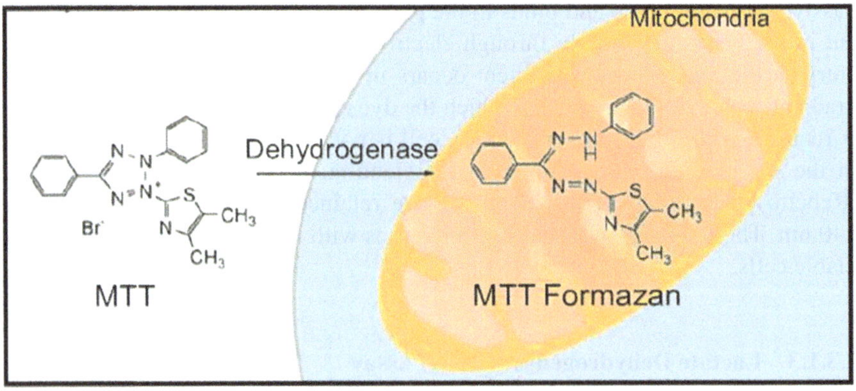


**MTT assay. (Courtesy: Stockert et al. 2018)**

The enzyme is capable of converting MTT tetrazolium salt to insoluble formazon crystals whose absorbance is at 540 nm. Viable cells produce more enzyme than the nonviable cells, thus there is more absorbance in viable cells. The colored formazon crystals do not give appropriate absorbance reading. In order to solubilize the crystals, there is a need for further steps and, thus, XTT assay is recommended.

#### 7.3.1.5 XTT Assay

XTT denotes the chemical name of the substrate sodium, 3′-[1-{phenylamino}-carbonyl]-3,4-tetrazolium]-bis(4-methoxy-6-nitro) benzene–sulfonic acid hydrate. XTT is slightly yellow colored when it is in the oxidized form, but the dehydrogenase enzyme reduces the compound and the color changes to bright yellow or orange color. The breakdown of the positively charged tetrazolium ring causes the color to change. The dehydrogenase enzyme is produced in the viable cells to maintain the pyridine nucleotide redox state (Manual).The absorbance was measured at 630–690 nm, and the absorbance increases with an increase in the number of viable cells.

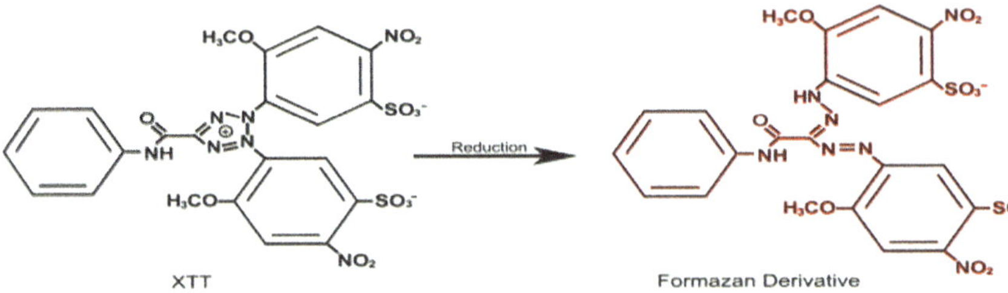


**XTT assay. (Courtesy: Manual)**

Since XTT is negatively charged and it cannot permeate the cell membrane, the reduction occurs on the outer surface of the cell with the help of trans membrane electron transport chains. This electron transport can further be facilitated by the chemical PMS (N- methyl dibenzine methyl sulphate). The reduced XTT is then taken in to the plasma membrane.

#### 7.3.1.6 WST (Water Soluble Tetrazolium) Assay

This test is very similar to XTT assay and it determines the viable cells by measuring the dehydrogenase enzyme, which is essential to maintain the redox state of the cell (Stockert et al. 2018).

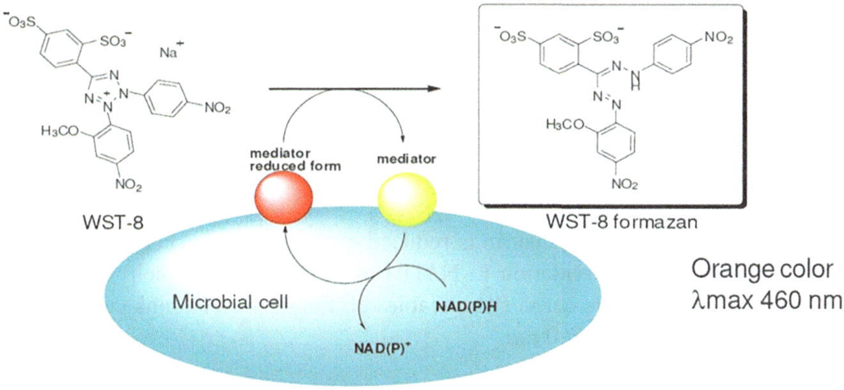


**Water soluble tetrazolium assay. (Courtesy: Stockert et al. 2018)**

The dehydrgenase enzyme in the presence of NADPH converts colorless WST −8 to a water-soluble orange colored formazan compound whose absorbance can be measured at 460 nm. Thus, absorbance increases with an increase in the number of viable cells.

#### 7.3.1.7 Tryphan Blue Assay

This assay determines the nonviable cells present in the culture. The basic principle is that the intact cells have plasma membrane such that, certain type of dyes cannot be internalized inside the cell, which include tryptaphan blue. Thus, viable cells containing an intact plasma membrane do not localize the dye and remain colorless, whereas dead cells which do not have an intact plasma membrane take the dye and become colored (Strober 2015).

#### 7.3.1.8 Live or Dead Cell Counting

This method is to measure or count the viable cells either by a spectrophotometer or haemocytometer (Riss et al. 2016). The detection is done by fluorescent-based detection or nonfluorescent-based detection.

Fluorescent-Based Detection

The viable cells present in the cell culture is detected by measuring the amount of esterase enzyme of the cell. The esterase enzyme converts the nonfluorescent substrate into a fluorescent compound. The fluorescence increases with increase in the number of viable cells. The fluorescent assays include the resazurin reduction assay concept, protease viability marker assay concept, and cell titer-glo assay test.

Resazurin Reduction Assay

In this test, nonfluorescent resazurin is reduced to resorufin by accepting the electron released due to the oxidation to NADH to NAD+ with the help of esterase enzyme produced in viable cells. Thus, viable cell fluoresces as a pink color whose absorbance is measured at 560 nm.

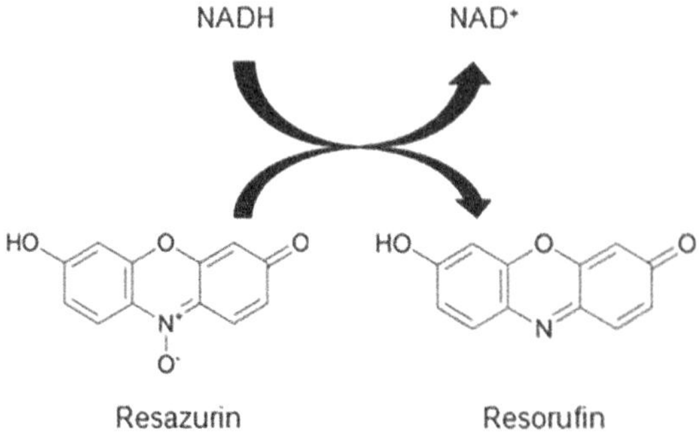


**Resazurin reduction assay. (Courtesy: Riss et al. 2016)**

Protease Viability Marker Assay

This is to assess the presence of viable cells by indirectly measuring the aminopeptidase enzyme, an active enzyme involved in protein synthesis. The aminopeptidase converts glycyl phenylalanile amino fluoro coumarin into aminofluorocoumarin whose fluorescent intensity is measured at 380–400 nm.

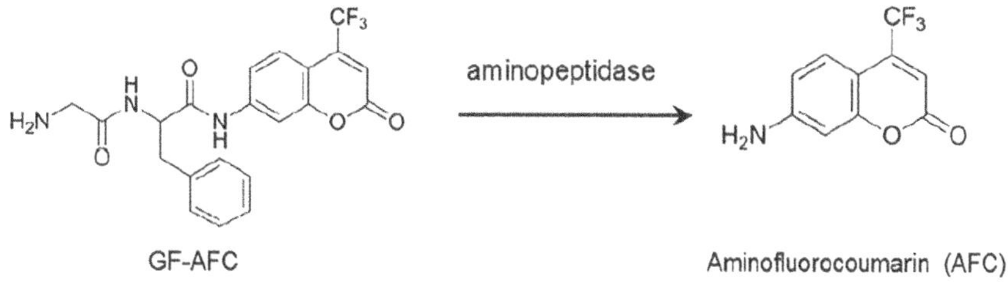


**Protease viability marker assay. (Courtesy: Riss et al. 2016)**

Nonfluorescent-Based Detection: Trypan Blue Test and MTT Assay

The nonfluorescent cytotoxic assays include tryphan blue exclusion, lactate dehydrogenase (LDH), tetrazolium assays (MTT, XTT, WST), and alamar blue assay (Dusinska et al. 2009). The optical density of the organic dyes used in the assays is measured. The color intensity often interferes with the dye and NP interactions, and eliminating such interference will be challenging. Therefore, 2D and 3D systems with clonogenic or plating efficiency assay (OECD test 476) are considered as the most reliable assays for cytotoxicity testing of nanofertilizers or any nanoproducts.

### *7.3.2 Genotoxicity Assay*

Genotoxicity test is done to determine the nanoparticles and cellular interactions that include DNA breaks (single or double stranded) mutations, deletions, chromosomal aberrations, micronuclei formation, DNA repair, and cell cycle. One of the most commonly used tests for genotoxicity is bacterial **Ames test** where DNA strand break measurements in cells are assessed using a comet assay, alkaline unwinding, and hydroxyapatite chromatography and alkaline elution. On the other hand, the cytogenetic assays are performed using micronucleus and chromosomal aberration tests, fluorescence in situ hybridization and chromosome painting.

Dusinska et al. (2012) have indicated that the genotoxicity for chemicals is determined by three different in vitro tests with two to three different end points. The chemical is considered genotoxic and carcinogenic when 1–2 tests were found positive and if any two tests are found negative, the substance is referred as "nongenotoxic." The Ames test (bacterial reverse mutation assay) is the most commonly used assay which is based on the induction of reverse mutation in the histidine gene that enables the bacteria to synthesize histidine and form visible colonies in minimal histidine medium.

The **comet assay** is a single cell gel electrophoresis which is one of the most commonly employed tests for exploring the genotoxic potential of NPs. It detects DNA damage in the cells that cause relaxation of supercoiled DNA, which then extends to form comet-like tails under electrophoresis. The relative intensity of DNA in the tail can be quantified that indicates the DNA break frequency. This method represents a rapid, sensitive, reliable, robust, and a relatively inexpensive way to study DNA damage (including DNA oxidation) and repair in different cell types, both in vitro as well as in vivo.

Genotoxicity of nanomaterials, micronucleus (in vitro and in vivo) mammalian gene mutation (in vitro), and Comet assays (in vitro and in vivo) are widely employed (Gonzalez et al. 2011; Doak et al. 2012; Pfuhler et al. 2013; Magdolenova et al. 2014).

The **genotoxicity assays** namely (1) Bacterial reverse mutation test, (2) Chromosomal aberration test, (3) Micronuclei test, (4) Cell gene mutation Test, (5) Comet assay, and (6) Unscheduled DNA synthesis are the commonly used tests.

#### 7.3.2.1 Reverse Mutation Test (Ames Test)

The Ames test is used for preliminary assessment of genotoxicity or point mutation. It utilizes different mutant strains of *Solmonella typhimurium* and *Escherichia coli* which can't grow in the absence of some essential compounds (i.e., minimal media) such as histidine for *S.typhimurium* and tryptophan for *E.coli*. The bacterial cultures used for the test are called mutants. When these mutant cultures are treated with nanoparticles, and if they are found to be grown in a minimal medium, then there are possibilities for the reversion of the mutation. Thus, cell culture, those grown after the nanoparticle's treatment, are called as reverse mutants. If reverse mutants are present, nanoparticles can be a genotoxic compound. The assay test can be performed according to the protocol of OECD **TG 471** (Dixit and Kumar 2018).

#### 7.3.2.2 Chromosomal Aberration Test

This test is to determine the chromosomal damage like break, deletion, addition, inversion, and transposition. It can be easily visualized during the metaphase stage of cell division, during which the chromatids will be arranged in the center of the cell before daughter cell formation. The overnight grown cultures are treated with nanoparticles, and the cells are harvested during the metaphase stage of the cell cycle. The cell cycle can be arrested at metaphase stage by adding colcemid chemical that helps in arresting the spindle fiber formation, which pulls the chromatids to opposite poles for cell division. Thus, metaphase stage easily visualizes the chromatids. The harvested cells are then incubated in dyes and observed under a microscope. The assay protocol can be followed as per the guidelines of OECD **TG 473** (No, 1997).

#### 7.3.2.3 Micronuclei Test

This test determines the presence of micronucleus in the cells formed as a result of clastogenic event, i.e., chromosomal damage. It can be easily visualized during the interphase stage of cell division. The cells can be arrested at interphase stage using cytchalasin B chemical. The harvested chemicals can then be stained and observed using a microscope. The assay protocol can be followed as per the guidelines of OECD **TG 487**(No, 2010).

#### 7.3.2.4 Cell Gene Mutation Test

This test is to determine the mutation at HPRT locus, which codes for the enzyme hypoxanthine guanine phosphoribosyl transferase gene, and XPRT locus, which codes for Xanthine guanine phosphoribosyl transferase. The enzyme HPRTase converts guanosine monophosphate to guanosine, an active enzyme essential in salvage pathway of nucleotide synthesis. In in vitro studies, the enzyme HPRTase converts 6-Thioguanine to 6-thioguanine monophosphate, 6-thioguanisine is a toxic compound, which inhibits cell growth. If there is any mutation in HPRT gene due to nanoparticles, it cannot convert 6-thioguanine to 6-thio guanine monophosphate and colonies will easily grow. Thus, the presence of colonies indicates genotoxicity of nanoparticles. The assay protocol can be followed as per the guidelines of OECD **TG 476** (OECD 2016).

#### 7.3.2.5 COMET Assay

Comet assay is also called as single cell gel electrophoresis assay. Till now, it is the most reliable test to identify the DNA break in individual eukaryotic cells. In this test, cells treated with nanoparticles and without nanoparticles were immobilized over the glass slide using agarose. The glass slide is now incubated in a lysis buffer to break the cell wall and plasma membrane without disturbing the double stranded DNA. Now, cells were treated alkaline buffer solution at which double strand DNA convert to single strand. Now the glass slide is stained with EtBr and subjected to gel electrophoresis and visualized under autoradiogram. If the DNA is damage due to nanoparticles, the autoradiograph will show a tail behind the spherical spot indicating the presence of small broken single strand DNA.

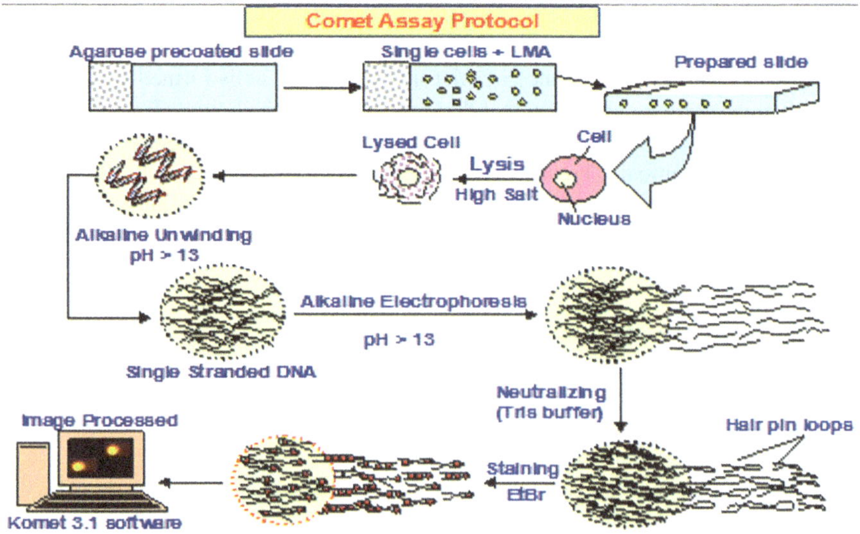


**Comet assay. (Courtesy: Andem et al. 2013)**

#### 7.3.2.6 Unscheduled DNA Synthesis

This determines the mutation or mismatches in the cell by utilizing the cell's natural nucleotide excision repair (NER). The NER is a DNA repair mechanism of normal dividing cells in which the damaged DNA or mismatched nucleotide, if present, will be replaced and ligated by the DNA polymerase I and ligases enzyme during replication. The cells treated with nanoparticles will be incubated in radioactively labeled thyamidine. If there is any damage in DNA due to nanoparticles, the cell's DNA repair mechanism utilizes radioactively labeled thyamidine to replace the DNA damage or mismatches. The amount of radioactive thyamidine consumed indicates the presence of mutation which can be found using an autoradiograph. The assay protocol can be followed as per the guidelines of OECD **TG 482**.

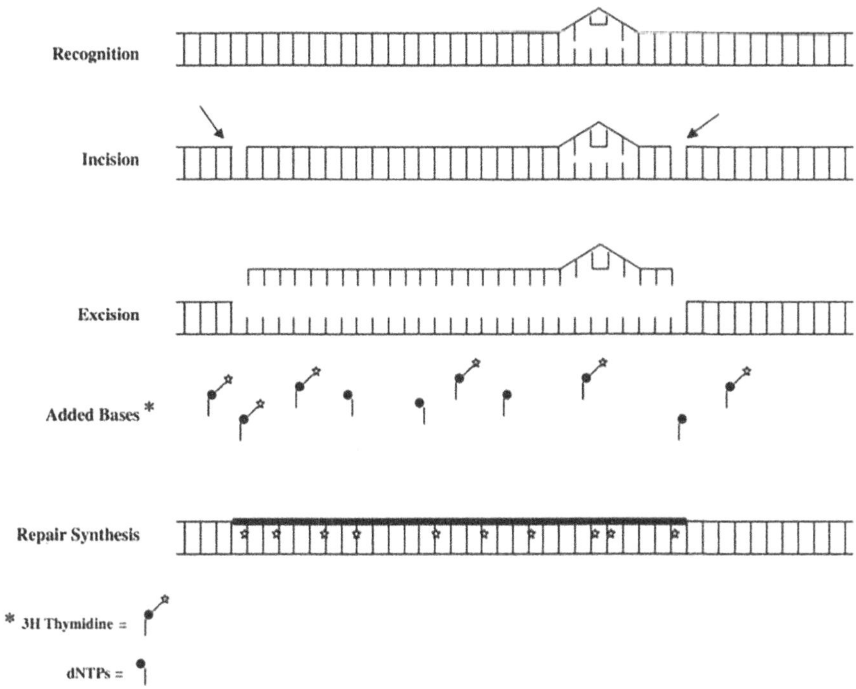


**Nucleotide excision repair. (Courtesy: Latimer and Kelly 2014)**

## 7.4 Risk Assessment and Management

Risk assessment is a complex process integrating hazard, exposure, and dose-response data to evolve a comprehensive system for the risk identification and management of nanomaterials. The risk characterization must be done by combining hazard (inherent toxicity of a compound) and exposure (concentration multiplied by the duration) to estimate risk (Hristozov et al. 2016). The "dose" refers to the quantity of NPs that enters a biological system or a specific organ (skin, lung, liver, etc.). The risk assessment management is quite complex, but it is universally accepted and adopted from UN reports that the risk assessment comprises four steps namely hazard characterization, dose-response determination, exposure assessment, and risk characterization.

### 7.4.1 Hazard

The first step in risk assessment and management of NMs is the hazard identification, which can be done by two approaches, namely weight of evidence (WoE) and Bayesian Networks (BN). The WoE is a statistical framework based on expert elicitation, while BN is a probabilistic, machine learning approach that can predict potential hazard of NMs. Among the two models, the BN exhibits more stability when the models are fed with new data (Sheehan et al. 2018). Thus, BN approach is gaining support in the context of NM risk assessment (Tsang et al. 2017), while the WoE has been criticized for being overly reliant on expert judgment and qualitative data.

### 7.4.2 *Dose-Response Relationship*

The dose-response assessment quantitatively determines the relationship between hazards and a concentration. Correlations between dose and biologically relevant endpoints or biomarkers are employed to determine no-observed adverse-effect levels (NOAELs) and human health exposure thresholds, such as recommended exposure limits (RELs) or occupational exposure limits (OELs) (Sheehan et al. 2018).

### 7.4.3 *Exposure*

The exposure will be assessed through one or more routes (i.e., inhalation, ingestion or dermal) or environmental compartments (e.g., water, sediment, soil) under varying exposure scenarios. Multiple-path particle dosimetry (MPPD) model is employed for the simulation of exposure to inhaled NMs in occupational and other scenarios, including animal experimentation (Cassee et al. 2002).

### 7.4.4 *Risk Characterization*

The hazard characterization, dose relationship, and the exposure assessment are done concurrently to gain insights of nature and magnitude of risks. The risk characterization includes health effects, evaluation of uncertainty and summarization of risk information and all the three steps are used to evolve tailor-made regulatory framework for risk assessment.

## 7.5 Risk Management Strategies

The Department of Science and Technology, Government of India, New Delhi, has brought out a document on "*Guidelines and Best Practices for safe handling of nano-materials in research laboratories and manufacturing industries*" that is widely used by the industries that are intended to produce nanofertilizers

### 7.5.1 *Exposure to Nanofertilizers and Recommended Safety Measures*

The safety measures to protect the workers in the nanofertilizer manufacturing plants and researchers in the QC laboratories are recommended to adhere to the safety measures. The exposure routes and safety measures are highlighted in Table 7.4.

**Table 7.4** Exposure, pathway entry in humans, and safety measures

| Type of exposure | Pathway | Safety measures |
|---|---|---|
| Dermal | Nanofertilizers may get through the skin and enter the body while handling solution or dry powder | Wear gloves and full sleeve lab coat |
| Ingestion | Nanofertilizers may be unintentionally absorbed and transported through the circulatory system | Eating and drinking are to be restricted in laboratories/manufacturing industries |
| Inhalation | Respiratory absorption by nostril lining, trachea, bronchioles, or the alveoli of the lung | The notified nanofertilizers in India are in liquid formulations that are handled with least risks. In the event of powder form of nanofertilizers in the years to come, use of respiratory air filters N100 or N95 are to be made mandatory |
| Injection | Skin puncture accidentally | Wear gloves and full sleeve lab coats |
| Ocular | Accidental entry of airborne droplets or particles of nanofertilizers into eyes, or transferred from hands while rubbing of the eyes | Safety glasses, goggles, full facepiece respirator are recommended. Avoid contact lenses in the manufacturing site |

### *7.5.2 Exposure Control Strategies*

In order to minimize the exposure of people to nanofertilizers, the following practices need to be in place: engineering and administrative controls, personal protective equipment, and waste disposal.

#### 7.5.2.1 Engineering Controls

Exposure to nanofertilizers should be minimized by the engineering control techniques, which include:

- Ventilation system affixed with high-efficiency particulate air (HEPA) filters to entrap the airborne NPs.
- Regular monitoring of ventilation and record keeping.
- Use gloves, fume hoods and exhaust enclosures when handling powders and pouring or mixing liquid media.
- Use fume hoods or other local exhaust devices.
- Air lock and sealed containers are to be used for collecting nanofertilizers from reactors.
- Evaporate the solvent from a colloidal dispersion using distillation under the explosion-proof enclosure.

#### 7.5.2.2 Administrative (Procedural) Controls

- **Training** is to be given periodically on SOPs, nanotoxicity, PPE, engineering controls and emergency procedures, waste handling, exposure monitoring and labeling and handling of nanofertilizer waste.
  - **Housekeeping:** Good housekeeping practices include clean working surfaces (i.e., work tables, glassware, equipment), regular vacuum cleaning and waste disposal as per hazardous-waste procedures.
  - **Work practices:** Ensure that material safety data sheet (MSDS) are available for all the ingredients used for the manufacturing of nanofertilizers. For ***liquid nanofertilizers***, wear gloves and avoid the dispersion of liquid droplets. ***For solid nanofertilizers***, handle it under the closed fume-hood or an enclosed vessel (glove box). In the event of sampling outside, wear class P3 (filters at least 99.95% of airborne particles) certified respiratory filters.
  - **Medical surveillance:** Health surveillance programs are a routine activity to monitor exposure levels. Some health issues like chronic immune responses of inflammation, allergy, respiratory disorders, gastrointestinal related disorder, and neurological disorders are to be checked.
  - **Record keeping:** Maintenance of records include training programs, risk assessments, testing of equipment, workplace monitoring, health surveillance (to be kept confidential), work related injuries and illnesses, work place engineering control maintenance, daily checks and examinations, and disposal records.
  - **Workplace monitoring:** Monitoring the particle concentrations using particle counters, condensation particle counters, scanning mobility particle sizers, and particle size analysis by electron microscopy.
  - **Storage:** Nanofertilizers are to be safeguarded to ensure workplace health and safety. Storage containers should accommodate different granulometric characteristics and reactivity of particles and are often needed to be protected from exposure to air and stored under inert gas/liquid bath.

#### 7.5.2.3 Personal Protective Equipment (PPE)

For those handling nanofertilizers, the following PPEs are to be strictly adhered to:

- Closed-toed shoes.
- Long pants.
- Long-sleeved shirt.
- Nitrile gloves.

- Goggles, safety glasses, and face shields.
- Respiratory air filters (N100 or N95).
- Full facepiece respirators.

#### 7.5.2.4 Waste Disposal

- Small quantities (in milligrams) of nanomaterials residues can be stored in containers, labeled and disposed as per stipulated SOPs.
- Quantities of nanomaterials (powders, colloids) including nano-fertilizers exceeding the milligram range should be treated as hazardous chemical waste, if the particle solubility in water is very small (inorganic materials like metals, metal oxides etc.).
- Small quantities of liquid nanofertilizer wastes can be treated with hydrochloric acid. Particles get agglomerated and precipitated and can be diluted and let out in a drainage dedicated for chemical disposal. The wastes must be collected, labeled, recorded, and disposed (incineration or recycling).

### *7.5.3 Best Practices to Be Followed while Handling Nanoparticles*

- Safety showers, fire extinguishers, and fire alarms are to be kept on red alert.
- Avoid eating food and beverages in nanotech laboratories or nanofertilizer manufacturing plants.
- Avoid mouth pipetting or siphoning.
- Rinse hands regularly and frequently to minimize potential chemical exposure through ingestion and dermal contact.
- Label all chemical containers with the term "nano" in descriptor (e.g., "nano urea or nano DAP").
- Wear gloves and work in a fume hood.
- Wear respiration protection when working outside a fume hood or in an open fume hood and consider overall protection.
- All contaminated material must be disposed of as chemical waste.

**Table 7.5** Specifications of notified nanofertilizers in India

| | | IFFCO | | CIL | Zuari | | TERI |
|---|---|---|---|---|---|---|---|
| No. | Specifications | Nano urea | Nano DAP | Nano DAP | Nano urea | Nano DAP | Nano P |
| 1 | N (%) | 1–5 | 7–9 | 2 | 8 | 6 | – |
| 2 | P as $P_2O_5$ (%) | – | 15–17 | 5 | – | 16 | 10 |
| 3 | pH | 4.5–6.0 | 3.0–6.0 | 4.5–6.0 | 7.0–8.0 | 7.0–8.0 | 6.0–7.0 |
| 4 | Specific gravity | – | – | 1–1.1 | – | – | – |
| 5 | Viscosity (cps) | 5–30 | 5–30 | 5–20 | 1.3–1.4 | 1.2–1.3 | 2–2.5 |
| 6 | Particle size (nm) (SEM/TEM) | 20–50 | 20–70 | 2–100 | 10–20 | 10–20 | 10–15 |

## 7.6 Labeling of Nanoproducts

Even though nanofertilizers are notified only after a complete biosafety test, response studies are conducted across 100 locations (minimum) and basic studies to justify the significance or uniqueness of the products. There are seven critical parameters to be labeled in bottles or containers. The key parameters include hydrodynamic particle size, physical size of the particles, pH, viscosity, specific gravity, nutrient concentrations and zeta potential (Table 7.5). These seven parameters are considered for quality assurance from the point of production until applied in the farmers' fields.

## 7.7 International Regulatory Framework

Nanofertilizers/nanoproducts are defined based on the particle size. In general, nanoparticles are defined as one-billionth of a meter or one-millionth of a millimeter. Nanofertilizers or nanoproducts should carry nanoparticles in a dimension of 1–100 nm. However, it is impossible to achieve 100% homogenized products carrying all particles <100 nm. A consensus has been arrived that a nano-product must have 50% of the particles in nano dimensdion of below 100 nm. The variations in the definition of nanoparticles across the globe is given in Tables 7.6 and The global scenario of the regulatory guidelines is presented in Table 7.7.

**Table 7.6** Nanoparticles and nanoproducts defined by various countries

| Country | Regulatory body/ responsible organization | Definition of nanomaterial: Size | Distribution threshold | Nanoscale dimension | Relevant legislation |
|---|---|---|---|---|---|
| European Union | European Commission (European Chemical Agency) | 1–100 nm | 50% by number | At least one | EC No. 1907/2006; Regulation, Evaluation and Authorization of Chemicals (REACH) for plant protection products and food contact materials |
| | *Specific to agri sector* European Parliament and the Council of the European Union on the Provision of Food Information to Consumers | 1–100 nm and can extend[a] up to 1 μm | Not mentioned | At least one | Regulation on the Provision of Food Information to Consumers (1169/2011) EC Regulation No 258/97 concerning "novel foods" and "novel food ingredients" |
| United States | United States Food and Drug Administration | 1–100 nm and extends[a] upto 1 μm | Not mentioned | At least one | Federal Food, Drug, and Cosmetic Act (FFDCA) |
| Canada | Health Canada | 1–100 nm and larger[a] | Not mentioned | At least one | Food and Drugs Act |
| India | Guidelines proposed by DBT | 1–100 nm and also 100-1000 nm if[b] | Not mentioned | At least one | Yet to make it legislate |

[a] Larger-size particles (up to 1 μm are considered as nanomaterials if it exhibits nanoscale properties)
[b] It has altered the agri-input or agri product characteristics associated with the application of nanotechnology compared with active ingredient
Adopted and modified from (Subramanian and Rajkishore 2018)

**Table 7.7** Global scenario of regulatory guidelines for nanoproducts in agri-food systems

| Country | Regulatory body/ responsible organization | Legislation | Provisions in the available legislation | Remarks |
|---|---|---|---|---|
| USA | Food and Drug Administration (FDA) and US-Environmental Protection Agency (US-EPA) | Federal Food, Drug and Cosmetic Act (FFDCA), Federal Insecticide, Fungicide and Rodenticide Act (FIFRA) | No specifications on nanoproducts in FFDCA or FIFRA | FDA recommends preliminary safety assessments at nanoscale but yet to develop regulatory guidelines |
| Canada | Canadian Food Inspection Agency (CFIA) and Public Health Agency of Canada (PHAC) | Food and Drugs Act | Nanoproducts are regulated under existing legislative | No nano-specific regulatory framework |
| EU | European commission (EC), European Parliament and Council | 1. Regulation on information to consumers (1169/20119)<br>2. Regulation on plastics in food contact materials (10/2011)<br>3. Regulation on active and intelligent packaging (450/2009)<br>4. The biocidal products regulation (528/2012)<br>5. The cosmetic products regulation (1223/2009)<br>6. Registration, Evaluation, Authorization and Restriction of Chemicals (REACH) regulation (1907/2006) | There are nano specific provisions<br>But no exclusive NM legislation | Taking a lead step in regulating nanoproducts |
| Switzerland | Federal Office for the Environment (FOEN)<br>Swiss Federal office of public health (FOPH) | | Existing regulations are ensured for safety of NMs | Physico-chemical parameters of NMs is mandatory to submit application for approval |

(continued)

Table 7.7 (continued)

| Country | Regulatory body/ responsible organization | Legislation | Provisions in the available legislation | Remarks |
|---|---|---|---|---|
| Australia and New Zealand | Food standards Australia New Zealand (FSANZ) | Australian New Zealand Food Standards Code | Amended FSANZ Application Handbook to support new food regulations to manage risks from nanoproducts | First regulatory program for "industrial nanomaterials was introduced by" National Industrial Chemicals Notification and Assessment Scheme (NICNAS) Tasmanian Institute of Agriculture has published a comprehensive review on Nanotechnology in Agricultural Products Logistics Management |
| China | Ministry of Agriculture Ministry of Health | Food Safety Law of China | National Centre for Nanoscience and Technology (NCNST) and the commission on nanotechnology standardization | Nanoproducts in agri-food systems is not yet approved by the authority |
| South Korea | Ministry of Food and Drug safety (MFDS), Korean Food and Drug Administration (KFDA) Korean Agency for Technology and Science (KATS) | National Nano-Safety Strategic Plan (2012/2016) and Food Sanitation Act | No NM specifications | Ministry of Knowledge and Economy (MKE) and the Korean Agency for Technology and Standards published guidance on safety management of nanobased products |
| Japan | Ministry of Health, Labour and Welfare | Food Sanitation Law | No NM specific legislation | Basic survey on safety of nanoproducts was done |

(continued)

**Table 7.7** (continued)

| Country | Regulatory body/ responsible organization | Legislation | Provisions in the available legislation | Remarks |
|---|---|---|---|---|
| Iran | Iran Nanotechnology Initiative Council (INIC) Nanotechnology Committee of Food and Drug Organization | | | Food and Drug Organization (FDO) constituted guidelines for nanoproducts in food, beverages, pharmaceutical, medical equipment, but agriculture is not yet included |
| Taiwan | Taiwan Nanotechnology Industry Development Association (TANIDA) | | Nanoproducts are certified | TANIDA established nanoMark system to certify nanoproducts |
| Thailand | Food and Drug Administration of the Ministry of Public Health | | | NanoQ label has been introduced for nanoproducts (but not for agri/food) that are certified by the Nanotechnology Association of Thailand |
| India | Food Safety and Standards Authority of India | Food Safety and Standards Act (2006) | | DST Nano mission developed guidelines for the safe handling of nanomaterials in the manufacturing industries<br>DBT has articulated the guidelines for evaluation of nano- agri inputs and nano-agri products in 2020<br>Based on the guidelines, nine nanofertilizers have been evaluated, notified, and commercialized for large scale adoption |

## 7.8 National Policy on the Use of Nanofertilizers in India

The Government of India constituted a committee comprising of Member of Parliament, ICAR Scientists, Experts from State Agricultural Universities and Representatives from Industries besides Secretary from the Ministry of Chemicals and Fertilizers. The Parliamentary Standing Committee on Chemicals and Fertilizers was chaired by one of the MPs, Dr. Shashi Tharoor, who presented its report on **'Nano-Fertilizers for Sustainable Crop Production and Maintaining Soil Health'** on March 21, 2023.

The Standing Committee brought down key observations and the following recommendations:

- Agriculture faces a bundle of challenges such as stagnation in crop yields, multi-nutrient deficiencies, and lower availability of arable land. The sustainable agricultural productivity is possible only through fertilizers as there is no scope for area expansion for cultivation. The imbalanced crop nutrition in India is primarily attributed to the excessive use of urea as a nitrogenous fertilizer.
- **Development of nanofertilizers:** IFFCO developed nano urea, which was notified in February 2021. It was the first nanofertilizer ever to be introduced in India or abroad. Subsequently, Nano DAP developed by both IFFCO and Coromandel International Ltd., were notified in 2023. Further, Zuari (nano urea and nano DAP) and TERI (Nano-P) got the nanofertilizers notified in 2023. So far, the Ministry of Fertilizers and Chemicals have notified nine nanofertilizers, which were brought under FCO regulations.
- **Benefits of nanofertilizers:** IFFCO indicated that a 500 mL bottle of nano urea (Rs. 240) can replace a 45 kg bag of conventional urea (Rs. 267) and that the nano urea costs less than the conventional urea. The nanofertilizers notified in India are liquid and recommended as foliar sprays. Overall, in the country, there was 8% increase in crop yields due to the introduction of nanofertilizers. The committee suggested that a dedicated long-term study is to be conducted to assess the benefits and side effects of nanofertilizers.
- **Challenges for adoption:** The cost savings from nano urea may assist in doubling the income of farmers. To motivate small and marginal farmers to adopt the technology, the Department has to undertake village level demonstrations and panel discussions on the radio to create awareness for nanofertilizers. Spraying is a drudgery for farmers and the committee recommended the Department to evolve cheaper spraying technology.
- There comes drone technology, which requires less time and labour. The cost of the kisan drones are Rs. 10 lakhs and the small and marginal farmers cannot afford them. Efforts should be initiated to conduct regular awareness programs and drone training to enable drone-based fertilizer sprayers.
- **Allocation of funds:** The Ministry should allocate a sizeable amount for nanotechnology-based research through fertilizer PSUs. It noted that development of varied nanofertilizers will help achieve self-reliance and save foreign exchange spent on increasing imports of fertilizers.

- **Imports:** Urea is a commonly used fertilizer by farmers in India and the subsidy increases every year along with imports to meet the growing demand. Judicious application of urea and nanofertilizers can reduce import dependency and self-reliance in the country besides saving of **Rs. 25,000 crores** in subsidies each year.

After the introduction of nano urea in 2021, there is a growing interest among fertilizer industries to introduce nanofertilizers. The Government of India is in support of introduction of several nanofertilizers that will likely substantially reduce the burden of fertilizer subsidies, which is running exorbitantly high at Rs. 2,25,000 crores for the year 2022–2023. The government has taken a policy decision that the introduction of nanofertilizers will serve as the plausible strategy to minimize the cost of import of fertilizers and the country is on the road to attain self-reliance.

## 7.9 FCO Regulations

The nanofertilizers in India are evaluated by an Expert Committee constituted by the Central Fertilizer Committee as per the stipulated DBT Guidelines for Evaluation of Nano based Agri-inputs and Food Products in India, which was launched in 2020 (Adholeya and Singh 2020). As per the guidelines, the first ever nanofertilizer, "Nano Urea," was notified in 2021. Under Schedule VII, clause 20 D of FCO, specifications were notified in respect of nano urea (liquid) fertilizer to be manufactured by IFFCO for a period of 3 years vide S.O.885 (E) dated 24th February 2021.

The Central Fertilizer Committee functions under the chairmanship of the Deputy Director General (Natural Resource Management), Indian Council of Agricultural Research, New Delhi. Expert members evaluate the nano fertilizer proposals for their safety as per DBT guidelines and send their recommendation to the Chairman of the Committee. Response studies are conducted in ICAR and SAUs across the country (at least 100 numbers). The committee also meticulously examines specific parameters such as the size of the particles (1–100 nm; hydrodynamic size using DLS and physical size by TEM), distribution of particles (50% of the particles <100 nm), zeta potential (−30 to +30 mV), viscosity, pH, specific gravity, and nutrient content. These seven parameters are notified by FCO and are often referred for the maintenance and monitoring of quality assurance at the farm gate.

## 7.10 Nanofertilizers Notified in India

### *7.10.1 IFFCO Nano Urea*

The IFFCO nano urea (liquid) used as foliar spray was the first ever notified nanofertilizer in 2020 by the FCO in India. The nano urea has 4% N. Foliar spray of nano urea (two rounds) may be equivalent or substitute for one top-dressing of conventional urea recommended for crops cultivated across the spectrum of crops. Further, it was observed that foliar application of nano urea along with basal application of the recommended dose of conventional urea has yield an advantage of 3%–8% over conventional fertilizer application. Nano urea appears to improve the growth and yield attributes (Midde et al. 2022).

### 7.10.2 IFFCO Nano DAP

IFFCO introduced yet another nanofertilizer, nano DAP (liquid), which was notified in 2023 by the FCO in India. The nano DAP has 8% N and 16% $P_2O_5$. Similar to nano urea, nano DAP can also be used as foliar spray (two rounds) instead of one top-dressing of conventional urea. The basal fertilizer remains unchanged and conventional fertilizer practice is to be adopted. In addition, foliar application of nano-DAP along with basal application of recommended dose of P has a yield advantage of 8–10% over conventional fertilizer application (Sahu et al. 2023).

### 7.10.3 Gromor Nano DAP

Coromandel International Ltd., Hyderabad, launched nano DAP (liquid) in June 2023 upon notification in 2023 by the FCO in India. The nano DAP has 2% N and 5% $P_2O_5$. The product is stable and well known to orchestrate physiological, biochemical, nutritional, growth and yield parameters. The studies conducted in ICAR institutes and SAUs have indicated that nano DAP as foliar spray (two rounds) helped to increase the yield of crops irrespective of agro-climatic zones. Conventional fertilizers of 75% and the remaining 25% of recommended dose P substitution through Gromor nano DAP (two rounds of foliar sprays at 20% and 40 DAS/DAT) is reported to produce 10–15% higher yields in comparison to conventional fertilizers.

### 7.10.4 Other NanoFertilizers

Recently, in the year 2023, four more nanofertilizers have been notified. The nanofertilizers include Zuari's nano urea (8% N), nano DAP (6% N; 16% $P_2O_5$), TERI's Nano-P (10% $P_2O_5$), and Ray's nano urea (4%). These fertilizers are notified but are yet to be made available at the farm gate.

## 7.11 Educational Programs

Among the 77 State Agricultural Universities in India, the Tamil Nadu Agricultural University, Coimbatore, is the first university to establish a state-of-the-art Center for Agricultural Nanotechnology in 2010. This is a unique facility encompassing a wide array of equipment for the synthesis, characterization, and applications, besides biosafety of nano-enabled processes and products. The Center is designed to undertake research on lab-on-chip technologies for on-site detection of pests, diseases, and nutrient deficiencies, nanobased agri inputs with higher use efficiency, preservation of perishables using nano interventions, besides biosafety. This is one of the best facilities in the country recognized by ICAR, GOI, and DBT as a referral laboratory for nanoproduct research, development, and evaluation. Despite the fact that nanotechnology education and research was primarily conducted by the Indian Institute of Science, IITs, CSIR laboratories, and Defence, TNAU was the first university to introduce UG subject (2008), PG (2011) and Ph.D. (2017) programs in Nano Science and Technology.

The TNAU served as a torchbearer in initiating an **UG course in Fundamentals and Applications of Nanotechnology** as early as in 2008 (TNAU Newsletter 2010). This had a remarkable impact across the country in implementing this starter course by other State Agricultural Universities. The TNAU has taught Nanotechnology courses to more than **46,000** UG students over the past 15 years. Six batches of 200 TNAU teachers were trained to teach the unique course. After several round of representations, **ICAR** recognized "Nanotechnology" subject as one of the UG mandatory courses entitled "GIS and Nanotechnology for Precision Agriculture" by the V Dean's committee.

The Tamil Nadu Agricultural University, Coimbatore, introduced a new PG program in Nano Science and Technology in 2011 (TNAU Newsletter 2011). The PG program is very **innovative and unique in the country,** wherein students are exposed to basics on nanotechnology, application in agriculture and allied sectors, nano-biotechnology and safety and regulatory framework. This helped TNAU to build **human resource** in the niche area of nano science, translating technique into technologies. More than **50 students** have acquired master's degree directly, and over **60 have used** nanotechnology as part of their research in the university till 2023. All the PG students have been financially supported through externally funded projects.

TNAU also introduced a **Ph.D. Program in Nano Science** and **Technology** in 2017 (TNAU News Letter 2017). Yet another **landmark** is the introduction of the Ph.D. program in Agricultural Nanotechnology in TNAU to nurture scholars who can employ and excel in their research and development**. Twenty Ph.D. scholars** have pursued their program and are in demand across the country as many SAUs identified nanotechnology as one of the priority areas. The scholars work on addressing agricultural field problems on **global challenges** and many are patentable or deserve to be published in high impact factor journals.

## 7.12 Concluding Remarks and a Way Forward

Research on nanofertilizers is recent, and for the past 10 years, this research has helped to develop a regulatory framework that has facilitated notification of nanofertilizers since 2021. While India introduced the first nanofertilizer in 2021, in the past 3 years, there has been tremendous support from the Government of India, which has resulted in the inclusion of eight more nanofertilizers in the country. Now, India serves as the torchbearer of nanofertilizers to the word and many Indian fertilizer companies have started exporting them to other countries, including the USA. The Government of India truly believes that nanofertilizers will pave the way to reduce the burden of fertilizer subsidy, which is currently Rs. 2,25,000 Crores annually. The data have clearly shown that there is a consistent response to the foliar feeding of nanofertilizers. Though there are critical views on nanofertilizers, the policy makers in the country truly believe that the technology will bring tangible benefits in the years to come. The lessons learnt have brought forth the following critical inputs for the nanofertilizer researcher to take forward.

- Basic studies on nanofertilizers must be undertaken to gain insights on entry, translocation, internalization, assimilation, and the fate of nanofertilizers in soil-water-plant-atmospheric continuum. This will resolve many paradoxical and unanswered quests.
- Nanofertilizers are highly soluble, reactive, and possess the ability to penetrate the cuticle, and nutrients are delivered to the target site. Specialty fertilizers are known to enhance crop growth, the yield, the quality, and the nutrient use efficiency, besides offering protection against abiotic and biotic stresses.
- Nanofertilizers available in the Indian market are liquid formulations that cause a wide array of practical difficulties in spraying, and the farmers prefer to have an easy mode of soil application. It is of utmost importance to develop nanofertilizers for soil to promote nanotechnology at the farm gate.
- Currently, nanofertilizers are intended to replace/substitute N and P, while other essential nutrients are yet to be developed as nano-formulations.
- Presently, there is no conclusive evidence to show that nanocarrier-based fertilizers boost nutrient use efficiency through improved nutrient transfer into the plant tissues when compared to conventional fertilizers.
- The soil scientist is still skeptical of nanofertilizers that carry nutrient contents in grams that are equated to kgs for conventional fertilizers. Nutrient balance studies, radio tracer studies, and the fate of nanofertilizers in the plant systems must be critically examined.
- The DBT Regulatory Guidelines for the evaluation and the use of nano-agri inputs in the country in a very generic framework and detailed protocols for various forms of nanofertilizers in relation to cropping systems will help gain a better understanding of the unique forms of fertilizers.
- Nano-fertilizers available in the Indian market are primarily derived from conventional urea or DAP. Nutrients are derived from these sources that are highly subsidized, converted into nano-sized particles/micelles through bottom-up

approaches (chemical/biological) and encapsulated with a polymer to achieve slow and smart release of nutrients.

- Futuristic nanofertilizers have to be designed and fabricated using agro-waste organic residues or industrial wastes through advanced molecular modeling similarly to drug designing.
- Nano-organic fertilizers can be thought of to cater to the current need of maintenance of soil organic carbon, which is critical to sustain farm productivity.
- There is a need to develop a comprehensive package of nanofertilizers carrying macro, micro, and trace elements to meet the balanced crop nutrition, while minimizing the cost of application.
- The fertilizer industries are to be encouraged to modify the machineries to suit manufacturing of customized nanofertilizer formulations.
- Though nanofertilizers are equated with conventional fertilizers with 90% subsidy in terms of costing, the new form of fertilizers may be given some support as an introductory benefit to the users.
- Nanofertilizer application may be encouraged through drone technology, which is gaining in popularity and is capable of extensive coverage.
- Nanotoxicity studies are to be undertaken extensively to determine the long-term effects of the use of nanofertilizers in the soil on beneficial microbes, micro fauna, and earthworms, and other associated effects.

Overall, the book chapter concludes that India has become a leader in nanofertilizers and the Government of India is looking at the nanofertilizer research as a ray of hope to reduce the burden of fertilizer subsidy while sustaining farm productivity. The regulatory framework for the evaluation of nanofertilizers is generic and more specific details are needed to gain insights into the products utilizing these cutting-edge technologies.

## References


Abd-Elsalam KA (2024) A comprehensive guide to nano-based fertilizers: types, production techniques, and properties. In: Nanofertilizer synthesis. Elsevier, pp 1–16

Abigail EA, Chidambaram R (2017) Nanotechnology in herbicide resistance. Nanostructured materials: fabrication to applications. IntechOpen, Rijeka, pp 207–212

Adholeya A, Singh R (2020) Guidelines for evaluation of nano-based agri-input and food products in India

Andem A, Agbor R, Ekpo I (2013) Review on comet assay: a reliable tool for assessing DNA damage in animal models. J Curr Res Sci 1(6):405

Anusuya P, Nagaraj R, Janavi GJ, Subramanian KS, Paliyath G, Subramanian J (2016) Pre-harvest sprays of hexanal formulation for extending retention and shelf-life of mango (Mangifera indica L.) fruits. Sci Hortic 211:231–240

Aruoja V, Dubourguier H-C, Kasemets K, Kahru A (2009) Toxicity of nanoparticles of CuO, ZnO and TiO2 to microalgae Pseudokirchneriella subcapitata. Sci Total Environ 407(4):1461–1468

Avila-Quezada GD, Ingle AP, Golińska P, Rai M (2022) Strategic applications of nano-fertilizers for sustainable agriculture: benefits and bottlenecks. Nanotechnol Rev 11(1):2123–2140

Baggs R, Ferin J, Oberdörster G (1997) Regression of pulmonary lesions produced by inhaled titanium dioxide in rats. Vet Pathol 34(6):592–597

Bansiwal AK, Rayalu SS, Labhasetwar NK, Juwarkar AA, Devotta S (2006) Surfactant-modified zeolite as a slow release fertilizer for phosphorus. J Agric Food Chem 54(13):4773–4779
Beig B, Niazi MBK, Jahan Z, Hussain A, Zia MH, Mehran MT (2020) Coating materials for slow release of nitrogen from urea fertilizer: a review. J Plant Nutr 43(10):1510–1533
Bernela M, Rani R, Malik P, Mukherjee TK (2021) Nanofertilizers: applications and future prospects. In: Nanotechnology. Jenny Stanford Publishing, pp 289–332
Cassee FR, Muijser H, Duistermaat E, Freijer JJ, Geerse KB, Marijnissen JC, Arts JH (2002) Particle size-dependent total mass deposition in lungs determines inhalation toxicity of cadmium chloride aerosols in rats. Application of a multiple path dosimetry model. Arch Toxicol 76:277–286
Chen Q, Qu Z, Li Z, Zhang Z, Ma G, Liu Z, Wang Y, Wu L, Fang F, Wei Z (2021) Coated diammonium phosphate combined with humic acid improves soil phosphorus availability and photosynthesis and the yield of maize. Front Plant Sci 12:759929
Dimkpa CO, Bindraban PS (2017) Nanofertilizers: new products for the industry? J Agric Food Chem 66(26):6462–6473
Dimkpa CO, Calder A, Britt DW, McLean JE, Anderson AJ (2011) Responses of a soil bacterium, Pseudomonas chlororaphis O6 to commercial metal oxide nanoparticles compared with responses to metal ions. Environ Pollut 159(7):1749–1756
Dixit M, Kumar A (2018) In vitro gene genotoxicity test methods. In: In vitro toxicology. Elsevier, pp 67–89
Doak S, Manshian B, Jenkins G, Singh N (2012) In vitro genotoxicity testing strategy for nanomaterials and the adaptation of current OECD guidelines. Mutat Res 745(1–2):104–111
Donaldson K, Aitken R, Tran L, Stone V, Duffin R, Forrest G, Alexander A (2006) Carbon nanotubes: a review of their properties in relation to pulmonary toxicology and workplace safety. Toxicol Sci 92(1):5–22
Dunn J, Fuller M, Zoeger J, Dobson J, Heller F, Hammann J, Caine E, Moskowitz BM (1995) Magnetic material in the human hippocampus. Brain Res Bull 36(2):149–153
Dusinska M, Dusinska M, Fjellsbø L, Magdolenova Z, Rinna A, Runden Pran E, Bartonova A, Heimstad E, Harju M, Tran L (2009) Testing strategies for the safety of nanoparticles used in medical applications. Nanomedicine 4(6):605–607
Dusinska M, Runden-Pran E, Carreira SC, Saunders M (2012) Critical evaluation of toxicity tests. Adverse effects of engineered nanomaterials, pp 63–83
El-Mahrouk E-SM, Atef EAM, Gabr MK, Aly MA, Głowacka A, Ahmed MAA (2024) Application of ZnO NPs, SiO2 NPs and date pollen extract as partial substitutes to nitrogen, phosphorus, and potassium fertilizers for sweet basil production. Plan Theory 13(2):172
FCO (2021) Fertilizer Control Order Published by the Ministry of Chemicals and Fertilizers, Government of India, Published by Fertilizer Association of India, New Delhi
FCO, 2023. Fertilizer Control Order, Published by the Ministry of Chemicals and Fertilizers, Government of India, New Delhi.
Fisher C, Rider AE, Han ZJ, Kumar S, Levchenko I, Ostrikov K (2012) Applications and nanotoxicity of carbon nanotubes and graphene in biomedicine. J Nanomater 2012:1–19
Forini MM, Pontes MS, Antunes DR, de Lima PH, Santos JS, Santiago EF, Grillo R (2022) Nano-enabled weed management in agriculture: from strategic design to enhanced herbicidal activity. Plant Nano Biol 1:100008
Ghafoor I, Habib-ur-Rahman M, Ali M, Afzal M, Ahmed W, Gaiser T, Ghaffar A (2021) Slow-release nitrogen fertilizers enhance growth, yield, NUE in wheat crop and reduce nitrogen losses under an arid environment. Environ Sci Pollut Res 28(32):43528–43543
Gil-Díaz M, García-Gonzalo P, Mancho C, Hernández L, Alonso J, Lobo M (2022) Response of spinach plants to different doses of two commercial nanofertilizers. Sci Hortic 301:111143
Gonzalez L, Sanderson B, Kirsch-Volders M (2011) Adaptations of the in vitro MN assay for the genotoxicity assessment of nanomaterials. Mutagenesis 26(1):185–191
Hannah R, Beck M, Moravec R, Riss T (2001) CellTiter-Glo™ luminescent cell viability assay: a sensitive and rapid method for determining cell viability. Promega Cell Notes 2:11–13

Hoecke KV, Quik JT, Mankiewicz-Boczek J, Schamphelaere KAD, Elsaesser A, Meeren PVD, Barnes C, McKerr G, Howard CV, Meent DVD (2009) Fate and effects of CeO2 nanoparticles in aquatic ecotoxicity tests. Environ Sci Technol 43(12):4537–4546
Hristozov D, Gottardo S, Semenzin E, Oomen A, Bos P, Peijnenburg W, van Tongeren M, Nowack B, Hunt N, Brunelli A (2016) Frameworks and tools for risk assessment of manufactured nanomaterials. Environ Int 95:36–53
Huang X, Wang X, Liu X, Cheng L, Pan J, Yang X (2024) Nanotechnology in agriculture: manganese ferrite nanoparticles as a micronutrient fertilizer for wheat. Plan Theory 13(10):1395
Jan NA, Subramanian K, Ganapathy S, Mohanraj J, Govindaraju K (2023) Nano-fiber enabled regulated release of hexanal vapor and its impact on shelf life of mango fruits. Polym Bull 80(1):865–881
Jędrszczyk E, Ambroszczyk AM (2016) The influence of NANO-GRO® organic stimulator on the yielding and fruit quality of field tomato (Lycopersicon esculentum Mill.). Folia Hortic 28(1):87–94
Jiang W, Mashayekhi H, Xing B (2009) Bacterial toxicity comparison between nano-and micro-scaled oxide particles. Environ Pollut 157(5):1619–1625
Jincy M, Djanaguiraman M, Jeyakumar P, Subramanian K, Jayasankar S, Paliyath G (2017) Inhibition of phospholipase D enzyme activity through hexanal leads to delayed mango (Mangifera indica L.) fruit ripening through changes in oxidants and antioxidant enzymes activity. Sci Hortic 218:316–325
Kah M, Kookana RS, Gogos A, Bucheli TD (2018) A critical evaluation of nanopesticides and nanofertilizers against their conventional analogues. Nat Nanotechnol 13(8):677–684
Khan M, Islam M, Nahar N, Al-Mamun M, Khan M, Matin M (2021) Synthesis and characterization of nanozeolite based composite fertilizer for sustainable release and use efficiency of nutrients. Heliyon 7(1):e06091
Kumar A, Saha S, Das A, Babu S, Layek J, Mishra VK, Singh R, Chowdhury S, Yadav D, Bhutia TL, Verma G (2022) Resource management for enhancing nutrient use efficiency in crops and cropping systems of rainfed Hill Ecosystems of the North-Eastern Region of India. Indian J Fertil 18(11):1090–1111
Kirschvink JL, Kobayashi-Kirschvink A, Woodford BJ (1992) Magnetite biomineralization in the human brain. Proc Natl Acad Sci 89(16):7683–7687
Lam C-W, James JT, McCluskey R, Hunter RL (2004) Pulmonary toxicity of single-wall carbon nanotubes in mice 7 and 90 days after intratracheal instillation. Toxicol Sci 77(1):126–134
Lateef A, Nazir R, Jamil N, Alam S, Shah R, Khan MN, Saleem M (2016) Synthesis and characterization of zeolite based nano–composite: An environment friendly slow release fertilizer. Microporous Mesoporous Mater 232:174–183
Latha M, Subramanian K, Sundara Sharmila DJ, Raja K, Rajkishore S, Chitdeshwari T (2023) Urea–lignin/chitosan nanocomposite as slow-release nanofertilizer, vol 3. ACS Agric Sci Technol, pp 463–476
Latimer JJ, Kelly CM (2014) Unscheduled DNA synthesis: the clinical and functional assay for global genomic DNA nucleotide excision repair. Mol Toxicol Protocols 1105:511–532
Lee WM, An YJ, Yoon H, Kweon HS (2008) Toxicity and bioavailability of copper nanoparticles to the terrestrial plants mung bean (Phaseolus radiatus) and wheat (Triticum aestivum): plant agar test for water-insoluble nanoparticles. Environ Toxicol Chem 27(9):1915–1921
Li H, Zhang J, Wang T, Luo W, Zhou Q, Jiang G (2008) Elemental selenium particles at nano-size (nano-Se) are more toxic to Medaka (Oryzias latipes) as a consequence of hyper-accumulation of selenium: a comparison with sodium selenite. Aquat Toxicol 89(4):251–256
Li H, Zhou Q, Wu Y, Fu J, Wang T, Jiang G (2009a) Effects of waterborne nano-iron on medaka (Oryzias latipes): antioxidant enzymatic activity, lipid peroxidation and histopathology. Ecotoxicol Environ Saf 72(3):684–692
Li X-b, Zheng H, Z-r Zhang M, Li Z-y H, Schluesener HJ, Li Y-y, S-q X (2009b) Glia activation induced by peripheral administration of aluminum oxide nanoparticles in rat brains. Nanomedicine 5(4):473–479

Lin D, Xing B (2007) Phytotoxicity of nanoparticles: inhibition of seed germination and root growth. Environ Pollut 150(2):243–250

Liu S, Wu Q, Sun X, Yue Y, Tubana B, Yang R, Cheng HN (2021) Novel alginate-cellulose nanofiber-poly (vinyl alcohol) hydrogels for carrying and delivering nitrogen, phosphorus and potassium chemicals. Int J Biol Macromol 172:330–340

Madlala NC, Khanyile N, Masenya A (2024) Examining the correlation of the nano-fertilizer physical properties and their impact on crop performance and nutrient uptake efficiency

Magdolenova Z, Collins A, Kumar A, Dhawan A, Stone V, Dusinska M (2014) Mechanisms of genotoxicity. A review of in vitro and in vivo studies with engineered nanoparticles. Nanotoxicology 8(3):233–278

Mahapatra O, Bhagat M, Gopalakrishnan C, Arunachalam KD (2008) Ultrafine dispersed CuO nanoparticles and their antibacterial activity. J Exp Nanosci 3(3):185–193

Manikandan A, Subramanian K (2014) Fabrication and characterisation of nanoporous zeolite based N fertilizer. Afr J Agric Res 9(2):276–284

Midde SK, Perumal MS, Murugan G, Sudhagar R, Mattepally VS, Bada MR (2022) Evaluation of nano urea on growth and yield attributes of rice (Oryza Sativa L.). Chem Sci Rev Lett 11(42):211–214

Mohan, C., Sridharan, S., Subramanian, K.S. Natarajan, N. and Nakeeran, S. 2017. Effect of Nanoemulsion of Hexanal on Honeybees. Journal of Entomology and Zoology Studies. 5(3): 1415–1418.

Mohan, C., Sowmya Priya, S., Sridharan, S. and Subramanian, K.S. 2020. Biosafety of Nanoemulsion of Hexanal on Chrysoperla zastrowi sillemi Stephens (Chrysopidae: Neuroptera)

Mohanraj J, Subramanian K, Rahale S (2022a) Enhancing multinutrients delivery and seed coating with biodegradable based electrospun nanofibre. Madras Agric J 109(1–3):1

Mohanraj J, Subramanian K, Raja K (2022b) Effect of multinutrients loaded electrospun PVA nanofibre on germination and its growth parameters of green gram [Vignaradiata (L.) Wilczek]. Legume Res An Int J 45(5):587–593

Navarro E, Baun A, Behra R, Hartmann NB, Filser J, Miao A-J, Quigg A, Santschi PH, Sigg L (2008) Environmental behavior and ecotoxicity of engineered nanoparticles to algae, plants, and fungi. Ecotoxicology 17:372–386

Nongbet A, Mishra AK, Mohanta YK, Mahanta S, Ray MK, Khan M, Baek K-H, Chakrabartty I (2022) Nanofertilizers: a smart and sustainable attribute to modern agriculture. Plan Theory 11(19):2587

OECD (Organisation for Economic Co-operation and Development) (2009) Environment Directorate Joint Meeting of the Chemicals Committee and the Working Party on Chemicals, Pesticides and Biotechnology

OECD (Organisation for Economic Co-operation and Development) (2016) Environment Directorate Joint Meeting of the Chemicals Committee and the Working Party on Chemicals, Pesticides and Biotechnology

Oberdörster G, Ferin J, Gelein R, Soderholm SC, Finkelstein J (1992) Role of the alveolar macrophage in lung injury: studies with ultrafine particles. Environ Health Perspect 97:193–199

Parhamifar L, Andersen H, Moghimi SM (2013) Lactate dehydrogenase assay for assessment of polycation cytotoxicity. Methods Mol Biol 948:13–22

Patel S, Beura S, Tripathy L, Dash A, Pandey G, Priyadarshani S, Madhusmita M (2020) Effect of different levels of Nanomax (4% NPK) and Polyfeed (19: 19: 19 NPK) on Coreopsis tinctoria. IJCS 8(5):1850–1853

Pfuhler S, Elespuru R, Aardema MJ, Doak SH, Maria Donner E, Honma M, Kirsch-Volders M, Landsiedel R, Manjanatha M, Singer T (2013) Genotoxicity of nanomaterials: refining strategies and tests for hazard identification. Environ Mol Mutagen 54(4):229–239

Phenrat T, Long TC, Lowry GV, Veronesi B (2009) Partial oxidation ("aging") and surface modification decrease the toxicity of nanosized zerovalent iron. Environ Sci Technol 43(1):195–200

Rajkishore S, Subramanian K, Natarajan N, Gunasekaran K (2013) Nanotoxicity at various trophic levels: a review. Bioscan 8(3):975–982

Rajkishore SK, Devadharshini KP, Sathya Moorthy P, Reddy Kiran Kalyan VS, Sunitha R, Prasanthrajan M, Maheswari M, Subramanian KS, Sakthivel N, Sakrabani R (2023) Novel synthesis of carbon dots from coconut wastes and its potential as water disinfectant. Sustain For 15(14):10924

Rajput VD, Singh A, Minkina T, Rawat S, Mandzhieva S, Sushkova S, Shuvaeva V, Nazarenko O, Rajput P, Komariah. (2021) Nano-enabled products: challenges and opportunities for sustainable agriculture. Plan Theory 10(12):2727

Reddy CH, Subramanian K (2015) Preparation and properties of chitosan coated nano dolomite for effective remediation acid soils. Trends Biosci 8:1009–1014

Repetto G, Del Peso A, Zurita JL (2008) Neutral red uptake assay for the estimation of cell viability/cytotoxicity. Nat Protoc 3(7):1125–1131

Riss TL, Moravec RA, Niles AL, Duellman S, Benink HA, Worzella TJ, Minor L (2016) Cell viability assays. Assay Guidance Manual [Internet]

Riss T, Niles A, Moravec R, Karassina N, Vidugiriene J (2019) Cytotoxicity assays: in vitro methods to measure dead cells. Assay Guidance Manual [Internet]

Sahu RK, Gangil S, Bhargav VK, Sahu P, Ghritalahre B (2023) Synthesizing biomass into nano carbon for use in high-performance supercapacitors-a brief critical review. J Energy Storage 72:108348

Selva Preetha P (2011) Nano-fertilizer formulation to achieve balanced nutrition in green gram. In: MSc.(Ag.) Thesis, Tamil Nadu Agricultural University, Coimbatore, India

Sheehan B, Murphy F, Mullins M, Furxhi I, Costa AL, Simeone FC, Mantecca P (2018) Hazard screening methods for nanomaterials: a comparative study. Int J Mol Sci 19(3):649

Shen CX, Zhang QF, Li J, Bi FC, Yao N (2010) Induction of programmed cell death in Arabidopsis and rice by single-wall carbon nanotubes. Am J Bot 97(10):1602–1609

Shoukat A, Saqib ZA, Akhtar J, Aslam Z, Pitann B, Hossain MS, Mühling KH (2024) Zinc and silicon nano-fertilizers influence ionomic and metabolite profiles in maize to overcome salt stress. Plants 13(9):1224

Shvedova AA, Kagan VE, Fadeel B (2010) Close encounters of the small kind: adverse effects of man-made materials interfacing with the nano-cosmos of biological systems. Annu Rev Pharmacol Toxicol 50:63–88

Singh A, Figueiredo R, Avagyan AA, Movsesyan HS, Rajput VD, Minkina TM, Singh RK, Sousa JR, Ghazaryan K (2024) Market value of nanotechnology-based products in agriculture: current status and future sustainability goals in the era of climate change

Sondi I, Salopek-Sondi B (2004) Silver nanoparticles as antimicrobial agent: a case study on E. coli as a model for gram-negative bacteria. J Colloid Interface Sci 275(1):177–182

Stockert JC, Horobin RW, Colombo LL, Blázquez-Castro A (2018) Tetrazolium salts and formazan products in cell biology: viability assessment, fluorescence imaging, and labeling perspectives. Acta Histochem 120(3):159–167

Strober W (2015) Trypan blue exclusion test of cell viability. Curr Protoc Immunol 111(1):A3. B. 1–A3. B. 3

Subramanian K, Rahale CS (2012) Ball milled nanosized zeolite loaded with zinc sulfate: a putative slow release Zn fertilizer. Int J Innovat Hortic 1(1):33–40

Subramanian KS, Rajkishore S (2018) Regulatory framework for nanomaterials in agri-food systems. Nanomaterials: ecotoxicity, safety, and public perception, pp 319–342

Subramanian KS, Manikandan A, Thirunavukkarasu M, Rahale CS (2015) Nano-fertilizers for balanced crop nutrition. Nanotechnologies in food and agriculture, pp 69–80

Subramanian KS, Muniraj I, Uthandi S (2016) Role of actinomycete-mediated nanosystem in agriculture. In: Plant growth promoting Actinobacteria: a new avenue for enhancing the productivity and soil fertility of grain legumes, pp 233–247

Subramanian K, Rajeswari R, Yuvaraj M, Pradeep D, Guna M, Yoganathan G (2022) Synthesis and characterization of nano-sulfur and its impact on growth, yield, and quality of sunflower (Helianthus annuus L.). Commun Soil Sci Plant Anal 53(20):2700–2709

Thirunavukkarasu M, Subramanian K (2014a) Surface modified nano-zeolite based sulphur fertilizer on growth and biochemical parameters of groundnut. Trends Biosci 7(7):565–568

Thirunavukkarasu M, Subramanian KS (2014b) Surface modified nano-zeolite used as carrier for slow release of sulphur. J Appl Nat Sci 6(1):19–26

Tian H, Liu Z, Zhang M, Guo Y, Zheng L, Li YC (2019) Biobased polyurethane, epoxy resin, and polyolefin wax composite coating for controlled-release fertilizer. ACS Appl Mater Interfaces 11(5):5380–5392

Tsang MP, Hristozov D, Zabeo A, Koivisto AJ, Jensen ACØ, Jensen KA, Pang C, Marcomini A, Sonnemann G (2017) Probabilistic risk assessment of emerging materials: case study of titanium dioxide nanoparticles. Nanotoxicology 11(4):558–568

Verma KK, Song X-P, Joshi A, Rajput VD, Singh M, Sharma A, Singh RK, Li D-M, Arora J, Minkina T (2022) Nanofertilizer possibilities for healthy soil, water, and food in future: an overview. Front Plant Sci 13:865048

Wang D, Saleh NB, Byro A, Zepp R, Sahle-Demessie E, Luxton TP, Ho KT, Burgess RM, Flury M, White JC, Su C (2022) Nano-enabled pesticides for sustainable agriculture and global food security. Nat Nanotechnol 17(4):347–360

Warheit DB, Laurence BR, Reed KL, Roach DH, Reynolds GA, Webb TR (2004) Comparative pulmonary toxicity assessment of single-wall carbon nanotubes in rats. Toxicol Sci 77(1):117–125

Wiwanitkit V, Sereemaspun A, Rojanathanes R (2009) Effect of gold nanoparticles on spermatozoa: the first world report. Fertil Steril 91(1):e7–e8

Yuvaraj M, Subramanian KS (2018) Development of slow release Zn fertilizer using nano-zeolite as carrier. J Plant Nutr 41(3):311–320

Yuvaraj M, Subramanian KS (2020) Novel slow release nanocomposite fertilizers. In: Nanotechnology and the environment. IntechOpen

Yuvaraj M, Subramanian KS, Cyriac J (2023) Efficiency of zinc oxide nanoparticles as controlled release nanofertilizer for rice (Oryza sativa L). J Plant Nutr 46(18):4477–4493

Yuvaraj M, Cyriac J, Subramanian KS, Hussien ET (2024) Slow release nanofertilizer for sustainable agriculture. In: Nanofertilizer delivery, effects and application methods. Elsevier, pp 53–69

The manufacturer's authorised representative in the EU is Springer Nature Customer Service Centre GmbH, Europaplatz 3, 69115 Heidelberg, Germany. If you have any concerns regarding our products, please contact ProductSafety@springernature.com

Printed and bound by CPI Group (UK) Ltd, Croydon, CR0 4YY
30/09/2026
02225309-0018